A Practical Guide to
Engineering Failure Investigation

A Practical Guide to
Engineering Failure Investigation

by

Clifford Matthews
BSc, CEng, MBA

**Professional
Engineering
Publishing**

Professional Engineering Publishing Limited
London and Bury St Edmunds, UK

First published 1998

© Clifford Matthews

ISBN 1 86058 086 6

A CIP catalogue record for this book is available from the British Library.

Typeset by David Gregson Associates, Beccles, Suffolk
Printed and bound in Great Britain by Bookcraft (UK) Ltd.

D
624.171
MAT.

Related Titles of Interest

Handbook of Mechanical Works Inspection	Clifford Matthews	1 86068 047 5
Process Machinery – Safety and Reliability	Edited by W Wong	1 86058 046 7
The Economic Management of Physical Assets	N W Hodges	0 85298 958 X
The Reliability of Mechanical Systems	J Davidson	0 85298 881 8
A Practical Guide to the Machinery Directive	H P Van Eklenburg *et al.*	0 85298 973 3
Industrial Sensors and Applications for Condition Monitoring	Phil Wild	0 85298 902 4
Assuring it's Safe	IMechE Conference 1998–6	1 86058 147 1
Plant Monitoring and Maintenance Routines	IMechE Seminar 1998–2	1 86058 087 4

For the full range of titles published by Professional Engineering Publishing contact:

Sales Department
Professional Engineering Publishing Limited
Northgate Avenue
Bury St Edmunds
Suffolk
IP32 6BW
UK

Tel: 01284 724384
Fax: 01284 718692

Contents

Acknowledgements

The author wishes to express his grateful thanks to the following people and companies for their assistance in compiling and reviewing the material in this book:

Tony Edwards *BSc(Hons)*
Principal Metallurgist, Royal and Sun Alliance Engineering (RSAE), Manchester, UK.

Martin C Yates
Principal Engineer, Pressure Equipment: RSAE.

John Lewis *BA(Hons), CEng, FIPlantE, ACII*
Claims Manager, RSAE.

Ian A Watt *ACII*
Manager, GAB Robins Ltd, London, UK.

Mike Wood
ERA, Leatherhead, UK.

John Freeman
Marine Surveyor and Consultant, Freeman and Partners Ltd, Fordingbridge, UK.

Bob Ross
Director, R. B. Ross Ltd, Glasgow, UK and author of *Investigating Mechanical Failures – the Metallurgists Approach*, 1995, Chapman and Hall.

Brian Hartley
Senior Inspection Engineer, Mott MacDonald Ltd, Brighton, UK.

Special thanks are due to Stephanie Evans for her excellent work in typing the manuscript for this book.

Preface

'We are here today, ladies and gentlemen, to hear why it failed – and I think we all understand what I mean by *it*' ... (guarded murmurs of expectations)

'Over to you Mr Failure Investigator' ...

'Thank you Madam Chairman, I'll get straight to the point. It involved fatigue, the 'old enemy',' ... (waits expectantly for the inevitable muted applause)

'Thank you for an excellent presentation, and the, er, source of this fatigue?'

'There are many possible sources of fatigue: vibration, wind loads, other cyclic stresses, and then there's bad surface finish, sharp corners, and of course sometimes there's ...'

'OK, but why did *this one* fail; what exactly was the *cause*?'

'Hmm, difficult to be specific. There are a number of possibilities that ...'

'He's got no clothes, he's got no clothes, the Emperor's got no clothes!'

'Ladies and gentlemen, order and decorum, *if* you please.'

Sooner or later most professional engineers will become involved in a failure investigation of some sort. Mechanical failures are perhaps the most common; even electrical equipment failures are often traced to mechanical components. Failure investigation is a difficult business (a rather awkward mix of technical disciplines, applied to a wide variety of different types of engineering equipment). It sometimes seems that, although the various theories of failure are well known, the answers in *your* particular failure somehow prove to be that little bit elusive. One of the main problems is the sheer variety of things that can cause engineering failure – coupled with the difficulties of understanding different materials and designs. You can see why this is a complex picture.

This is not another book on metallurgy – there are plenty of excellent ones on the market already. I recommend that you read them: they hold core technical information that you will need for failure investigation. They do not, however, contain *all* the information that you need to show you how to conduct a failure investigation. I have learnt that they only give you part of the picture. As well as pure technical information you need to know something about the procedures and techniques of the investigation process – about its *pattern*. You also need to have a firm idea of the methodology and structure of what you are doing, to help slash the variety of the subject down to something that you can handle.

There are two parts to this book. Part I deals with the techniques and Part II the more technical parts, with specific aspects related to insurance and commercial investigations. Both parts provide different 'technical pointers' to performing failure investigations that are effective, and that have some real sales value. If you are an advocate of the fudged or hedged conclusion, or favour the well-worn principle of drawing no conclusions at all, then this book is, unashamedly, not for you. Please turn, with my best wishes, to the bibliography at the back.

Finally, in deference to experienced failure investigators (and Emperors), I am open to criticism of the content of this book. If you find errors or inconsistencies (and which of us is perfect?) then please let me know. You can write to me c/o The Publishers, Professional Engineering Publishing Limited, at the address given in the front of the book.

Clifford Matthews BSc, CEng, MBA

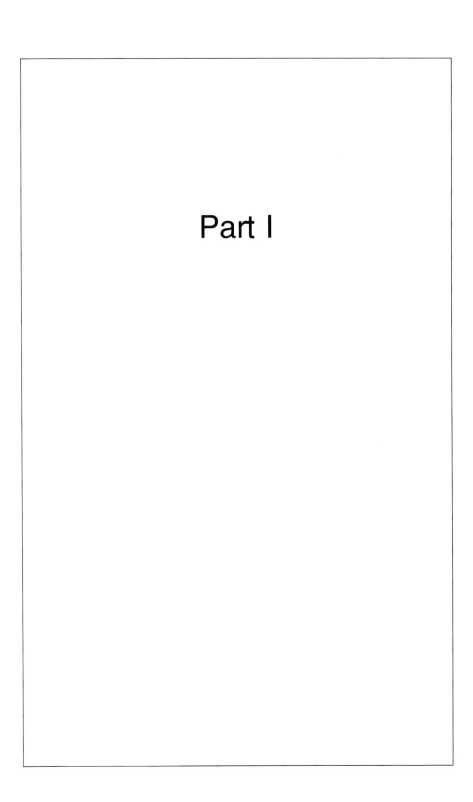

Part I

Chapter 1

About this book

What is so special about engineering failure investigation? Isn't it easy to look at the technical facts, listen to the explanations of what happened, and then decide what caused the failure? Unfortunately, this is not so, indeed it often seems as if the opposite is true:

MANY FAILURE INVESTIGATIONS END UP IN A MESS.

Of those that don't, the investigations themselves frequently become long and inconclusive affairs, and lead to poor or contrived answers. Others never really fulfil their objectives, or get to the core of the failure. Perhaps the most dangerous category – but a very real one, nevertheless – are those investigations that degenerate into a search for consensus, rather than an inquiry into the real reasons for failure.

So why is this? I am sure that there are several reasons. One of the most important ones is that pure engineering knowledge, around which failure investigations revolve, is in reality only a component part of the investigation process. It is surrounded by the *context* of the failure and tempered by the way that engineering conclusions can be proved or disproved – a complex picture indeed.

This book is a basic guide to investigating failures of engineering equipment. It should be relevant to all plant failures that have a *mechanical* basis to them, so it does not exclude, for instance, items of electrical equipment or even civil engineering structures that have suffered failure. It is intended for engineers who may become involved in a wide variety of failure cases, extending across technical disciplines. I have tried to make it *useful*, with a practical approach, rather than filling it with technical definitions, or (yet more) metallurgical theory. There are plenty of (other) authoritative books on metallurgy. What this book will do is to widen your appreciation of failure investigations outside your engineering discipline; that is, outside what is probably your technical comfort zone. I make no apologies for this; there is probably little choice because if you want to become competent at failure investigation, you will soon run into the fact that your engineering knowledge is only part of the story.

Failure investigations – for what purpose?

Another precept of this book is that every failure investigation must have a purpose and then, more importantly, it must have a *point*. Sadly, so many of them don't. Being effective at failure investigation is about two things: *finding* the point, and then *keeping* to the point. This is easier than it sounds. Figure 1.1 shows how the purpose of any failure investigation can be apportioned into one or more of three types. Note the top one on the list – more than half of all failure investigations are carried out as part of an insurance claim, either by the claimant or, indirectly, by the insurers and their representatives. Insurance, of course, is about *money* and the core purpose of the technical investigation is normally related to deciding who pays for the consequences of the failure. The second type identifies commercial liability as the major purpose of the investigation and you will not be surprised to find out that this is ultimately also about money. Approximately 35 percent of engineering failure investigations are undertaken with the objective of finding who was responsible for the failure. This means the process involves apportioning *blame*. You could be forgiven for thinking that the third possible purpose, technical improvement, is in a minority category. This has some justification: there are precious few liability- or insurance-driven investigations that look for ways to improve the failed item in preference to identifying what caused the failure or who was to blame. It is also true, however, that even in the purest and most divisive of liability cases, real engineering improvements do follow from the

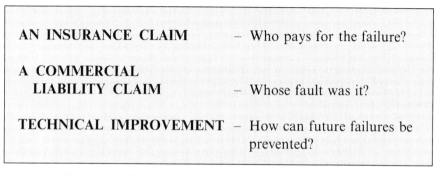

AN INSURANCE CLAIM   – Who pays for the failure?

A COMMERCIAL LIABILITY CLAIM   – Whose fault was it?

TECHNICAL IMPROVEMENT – How can future failures be prevented?

Figure 1.1   Three possible purposes of a failure investigation

results of the technical failure investigation. There is no real paradox here – tough incisive conclusions mean better technical improvements. Try to keep a clear focus on Fig. 1.1. Its content will be reflected in several forthcoming chapters.

An effective approach

Together with purpose goes *approach*. The general methodology for failure investigation put forward in this book forms the 'key' to the structure of its chapters. It is not a particularly novel approach and I am almost sure that it is imperfect. Its only virtue is that it is structured, and so can help you bring some organization to a failure investigation. Figure 1.2 shows the outline of the methodology: you will see this developed throughout the technical chapters of the book and (hope-

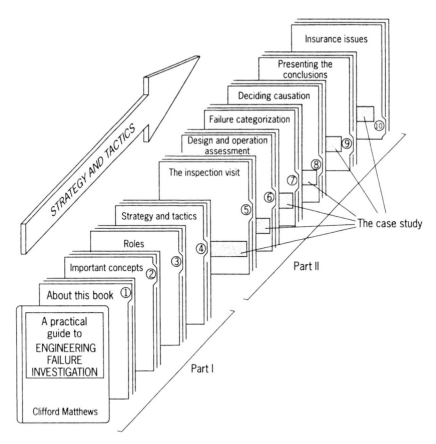

Figure 1.2   This book – its methodology for failure investigation

fully) understand how it follows the general 'pattern' that failure investigations take. The way in which you think about failure investigations is an important part of the correct approach. We will look at some techniques of 'clear thinking' and 'logical progression' as the methodology unfolds.

Using the book

You will see that there are two separate parts to this book. Part I (Chapters 1 to 4) covers the essential background to failure investigations. Chapter 2 introduces the important concepts (and a little formal terminology) that you need to know. Some of these are engineering-based but there are also those related to some of the insurance- and liability-related aspects of failures (remember what I said earlier about your having to take short steps outside your *comfort zone*?). Chapter 3 is about roles, so it has to discuss the parties involved in a failure investigation. 'Parties' means *people*, and their opinions – of engineering, and maybe of you (how is your 'comfort zone'?). Part I ends with Chapter 4, which contains guidance on strategy and tactics. Don't expect this to be anything to do with the general 'management' philosophies or definitions that sometimes surround these terms – it is about the strategies and tactics of failure investigations only, nothing else. Chapter 4 is intended to help your understanding of what happens during a failure investigation: it will not make you into a manager.

Part II (comprising Chapters 5 to 10) tracks the activities of the failure investigation itself. All the chapters have a strong technical root and follow, broadly, in the chronological order of a real failure investigation. Parts I and II are each intended to be consistent in themselves, so:

THERE IS NO REASON WHY YOU HAVE TO READ PART I FIRST.

By reading Part II first, you can see the methodology in action, in an actual case study, *before* reading the justification behind it. Sometimes this is a better way to see the overall picture. You decide, but be careful not to skip about too much *within* Parts I or II, or you might lose sight of the pattern that failure investigations follow. The case study develops throughout the chapters: it is about the failure of a straightforward piece of rotating machinery, a large radial fan. There is nothing particularly special or new about the technical content of this case study – its purpose is to help crystallize the ideas from Parts I and II, by showing them in action.

A reminder – your product

Finally, in this introductory chapter, a gentle reminder again of the purpose of failure investigations – of the *product*. As a failure investigator, you will need to produce a result. So, you have to:

- *investigate* the failure
- *decide* the cause; 'causation'
- *present* causation.

On balance, it is the last two that hold the intrinsic commercial value. The actual process of investigation, by itself, is worthless unless it leads to definitive answers. This means you have to look for *clear conclusions* to engineering failure investigations, not heavily qualified half-answers. The objective is to aim for precision and avoid procrastination; and remember the problems of going for the easy-option 'consensus' approach which almost never has much real commercial value. Keep on doing it and your credibility in engineering failure investigation will slowly drain away.

Chapter 2

Important concepts

Failure investigations involve the use of a set of specific terminology and concepts. Some of the terminology used appears in its literal sense, so it has the same meaning as when used in a general technical discussion, but there are a few terms which have specific meanings in the context of failure cases. We can look at these in turn.

Types of investigation

Figure 1.1 showed that there are essentially three *possible* purposes of investigation: we can now expand these to take into account the slightly wider variety that can be encountered in real engineering situations. Figure 2.1 is a diagrammatic view of four *types* of investigation designated, for convenience only, as A, B, C, and D. Note how all are shown as having a common technical 'core'. The types have different characteristics, as described below.

Type A – Commercial investigations
There are really three separate subtypes in here: contract-related investigations, those with a purely commercial basis, and insurance claims.

- A typical example of a *contract-related* investigation is the case in which a pump, installed in a large process plant, fails during its guarantee period. The main question here is to identify the failure mechanism so it can be decided whether the failure was caused by the pump's user, by actions which may invalidate the warranty awarded under the purchase contract.
- Pure *commercially based* failure investigations are similar to that above but not so shrouded in specific contract conditions – although there may be a general contract, such as the Sale of Goods Act, in the background. A typical example is the motor vehicle camshaft that

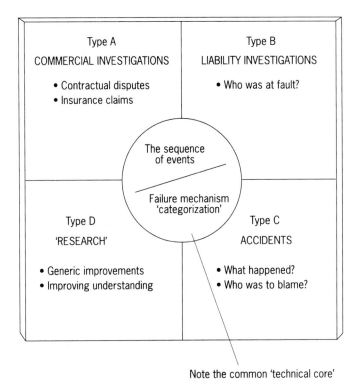

Note the common 'technical core'

Figure 2.1   Four 'types' of failure investigation

fails after 20 000 km of use, but outside the vehicle's contracted guarantee period. The prime purpose of this type of investigation is to conclude whether the component was 'fit for purpose' and of 'merchantable quality' – common Sale of Goods Act terminology.

• *Insurance cases* are related to the key subjects of the failure *event* and its *causation* – again, formal terminology. Insurance-related failure investigations do have the advantage that the contractual framework is clearly defined (in the form of the insurance policy wording) – which is surprisingly simple, at least in the way that it impinges on the interpretation of the technical conclusions of a failure investigation.

Type B – Liability investigations
There is a fine line between type A and B investigations. The differences lie in the depth of the investigation and in the precise form that the conclusions take. Liability cases are centred around technical evidence of the *actions* (or the lack of them) that are taken by the involved

parties. Consider, for example, a diesel engine that suffers from bearing failure because of lubricating oil starvation or contamination. Stated like this, this technical conclusion may well be adequate for the purposes of an insurance claim, but to apportion properly liability for the failure it is necessary to follow the trail further backwards in time to ask *why* the oil starvation or contamination occurred. What actions led to this happening, or what should have been done, that wasn't? This makes liability cases difficult – you sometimes have to be very precise in your findings, in order to be able to apportion technical blame.

Type C – Accident investigations

This is always the most controversial category, but not necessarily the most difficult. These investigations are similar to liability cases in that they have to decide *who* was at fault. There is also, however, the aspect of improvement and future accident prevention feedback to add to the picture – this is a common feature of even the smallest accident investigation. Failures of statutory equipment such as cranes and pressure vessels provide good examples: the investigation procedure has to point out who was responsible under, for instance, the provision of the Health and Safety at Work Acts, but also to address the wider lessons of what can be learnt from the failure. Such findings are used to amend design procedures and the technical standards referenced in the relevant statutory instruments.

Type D – 'Research' investigations

Perhaps this is a slightly unfair title, but that is essentially what they are – engineering investigations in the purest sense, often with a distinct scientific or even academic trait. You may see them called 'design studies' or 'surveys' – their common features tend to be the collection, over time, of diverse failure data from different sources. The investigations rarely happen quickly, or result in dramatic conclusions – you can think of them as adding to the overall subject of engineering failure investigation, but without being too case-specific. They are not addressed specifically in this book, but are part of the picture, nevertheless.

Figure 2.1 summarized the major characteristics of type A, B, C, and D failure investigations. There are real differences between these types and it is important to treat each type in its own way, to get the most effective results.

Core content

Differences in objectives and detail apart, some of the core content of type A, B, C, and D investigations is *common* to all of them. As a rule of thumb, about half of the 'technical content' can be considered common, whichever type of investigation a particular failure case belongs to. There are two essential *parts* of this core content, both of which are important concepts of failure investigations: the sequence of events and failure mechanism *categorization*. These terms have specific meanings – we will look at them in turn.

Sequence of events

The concept of *timescale* is important in failure investigations. The order in which things happened (the proper term is *events*) has a significant effect on the way that the conclusions are drawn, and on the eventual result of the investigation. Failures never 'just happen', there is always a series of events which occur over time.

There are a few practical points to be considered here – you can think of these as constraints on the description of the sequence of events:

- The description of events must be linked to a *clear* timescale.
- The *sequence* of events has to be explained.
- Any descriptions must be strictly *technical*, describing what happened, not why it happened (this comes later).

Failure mechanism categorization

Basically, 'categorization' involves conclusions which are drawn about the technical 'happenings' of a failure. They are *absolutely* technical, and do not refer, at all, to what caused the failure. The issue of *causation* (another formal term) will need to be addressed, but the description of failure categorization is not the correct place within which to do it. Failure categorizations will later be used to *lead to* the conclusions of the failure investigation, however, so they must be clear. They must also be technically consistent – you can think of them as 'lowest common denominator' technical conclusions, which will be added to (increased in complexity) as the investigation progresses. The concept of categorization, and how it fits into the overall scope of the failure investigation, is discussed more fully in Chapter 7.

The damage/causation framework 'model'

The easiest way to understand the subject of failure investigations is to use a framework, or 'model', to help explain what is happening. Be careful not to confuse this with a *methodology*: the framework is more an outline of the elements of a failure case, rather than a recommended method of *conducting* the investigation process.

The basic framework is shown in Fig. 2.2. It is not a perfect model: its strength lies in the fact that it is applicable to all four types (A, B, C, and

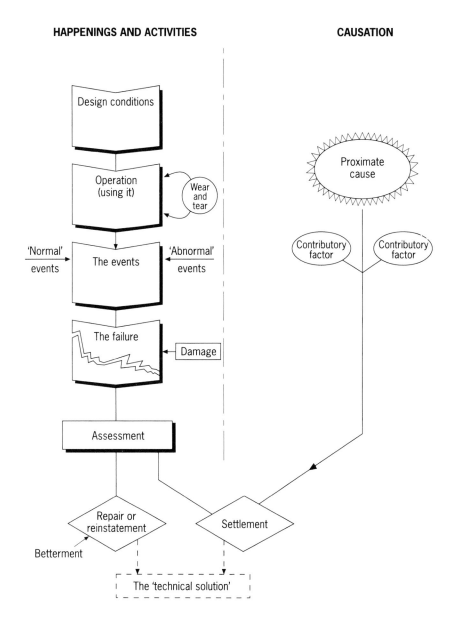

Figure 2.2   The framework of damage and causation

D) of failure investigation. Its purpose is to be *useful*: to help your understanding of the different elements and definitions, and how they fit together. Note how there are two 'sides' to the framework: the left hand side (LHS) is about *findings* and *events* – things that actually happened – while the right hand side (RHS) is more about *opinions* – the main category of opinion being that related to causation. The vertical axis is a broad representation of timescale, the activities proceeding more or less chronologically from top to bottom. Note how the top areas of the framework refer to the period when the plant or component was operating in its pre-failure condition. It ends, at the lower end, after all the post-failure activities have been completed, including repairing the failure (the formal term is 'reinstatement') and agreeing a formal settlement on the payment and liability aspects. Before preparing further to use this damage/causation framework (you will see it being used in Part II of the book) we need to look at the definitions of some of the major elements.

Design conditions

These are, simply, the conditions for which an engineering component is designed. They range from general conditions of duty, speed and ambient temperature range to more complex technical parameters such as specific loadings, bending moments, fatigue limits and the like. Design conditions are not always stated explicitly – often they have to be inferred by general technical engineering knowledge (or sometimes only experience).

Operation

This is the period when the component is being *used* for the purpose for which it is intended. Preliminary activities such as commissioning and testing are normally not considered as part of the operation period. One of the consequences of the operation period is that it causes *wear and tear* (again, a formal definition that you will see in common use) to the component. This starts to use up its design *lifetime*.

Events

'Event' is a widely used term in failure investigation; one which can be used in several ways. An event is:

- Something that *happens*.
- *Identifiable*: i.e. the event itself can be identified, even if it is hidden among lots of other events.
- An activity that causes a *result*.

Note the precision of the last two definitions – these are not just semantics, you will see later how these definitions hold significant importance when it is necessary to start defining conclusions, particularly in type A (insurance-related) failure investigation. There are three main categories of events.

- *Normal events* are events that can be considered unremarkable in comparison to the planned operating condition of a piece of engineering equipment. They are within the design conditions intended for the equipment assembly, and all its component parts. Note that the term 'normal' does not exclude the possibility of external 'interference' in the operating regime of a component – a motor vehicle, for example, is designed to provide some resistance to crashes, so a two-vehicle collision would not necessarily be classed as an abnormal event. In the context of failure investigations, 'normal' really means *expected.*

- *Abnormal events* are the opposite to normal events. An abnormal event is one which is, by definition, unexpected – it is not part of the original design concept of the component. Abnormal events are important in insurance-related (type A) failure investigations.

- *Failure events* are slightly more difficult to understand. The definition of a 'failure event' is a situation in which an event occurs that provides the *conditions* for failure to occur. It does not actually have to initiate the failure at the precise moment of time that 'the event' is considered to occur, although in most cases it does. As an example, a ship's propeller hitting a submerged object can be classed as a failure event – the propeller blades may break immediately after the impact, or later, having been weakened. It is the *impact* that is classed as the failure event, not the action of the blades breaking off. Failure events may happen singly, or in unbroken chains. The concept of *an unbroken chain of events* is useful – we will return to this in Part II of the book.

Failure

The term 'failure' provides a good example of the difficulties that you can encounter when working in engineering failure investigation. The prime point is:

'FAILURE' CAN HAVE TWO DIFFERENT MEANINGS – DEPENDING ON THE CONTEXT.

The 'plain-language' definition, first, is the most obvious: 'failure' means something is bent, twisted, burnt or broken, so that it is in some way unable to do the job for which it was intended ('designed'). Normally, the failure is the result of a failure *event*, as we discussed earlier. The strict mechanical engineering definition is, however, different – here, 'failure' is the general term used for a condition in which an engineering member exhibits plastic deformation. Once plastic deformation occurs, any deformation ('strain') that has occurred is, by definition, irreversible. The difficulty with this engineering definition is that a component that is formally in a 'state of failure' may actually be in one of several conditions: damaged, fractured or broken (ruptured). Again, we will look at these later in Part II of the book – the main point at the moment is to understand that 'failure' has different plain language and formal engineering definitions. This means that you need to be extremely careful with this term, *every* time you use it.

Damage

Like 'failure', the term *damage* has several meanings, depending on the context in which it is used. The plain language and engineering definitions of 'damage' are sometimes so different that they can actually look like opposites. In common usage the term 'damage' normally means that the item is broken, or in some way degraded, to the extent that it cannot be used. This definition is weak. Returning to the analogy of the motor vehicle, a collision would normally be described as resulting in bodywork 'damage', although there is no indication in this term whether the damage is superficial, or serious enough to make the vehicle unusable. The other feature of common usage is that damage is usually considered to be caused by an *event* (in this case the vehicle collision) rather than being the result of normal operation.

The engineering definition of damage is different. In this context, 'damage' is not necessarily the result of an event (although it *could* be): it is defined as the condition in which some plastic deformation has occurred, relative to the 'as new' condition of the component. The key point about this definition is that it assumes that the damaged component can still be used. So:

Damage does not mean *failure*

In the broad sense, 'damage' can include the effects of fatigue, wear or corrosion, so it may be visible, or not. Wear and corrosion are usually visible, macroscopically, whereas fatigue effects are not. Now for the complicated part: bearing in mind that there are two (almost opposing) generic definitions of 'damage', we need to move on to look at the specific *ways* that damage is referred to in failure investigations. These are predominantly non-technical ways of talking about damage, but they do form a very real part of the 'territory' of commercial insurance and liability (type A and B) investigations. This means that you will be expected to use them in your conclusions – so it is wise to try and gain an understanding of them. There are four main concepts:

- direct damage
- consequential damage
- indirect damage
- related damage.

It is only really possible to get the full picture of these definitions by looking in some detail at an actual failure investigation – this is covered in Part II. For the moment, as an introduction, we will look at a highly simplified failure example.

> *Scenario*: The stranded steel rope of an overhead crane has stretched then snapped when the crane attempted to lift more than its designed safe working load (SWL). The load consisted of heavy but fragile castings which broke when they hit the shop floor. The shop floor is also damaged. The broken rope hit and broke some of the crane's electrical micro-switches.

Direct damage

Direct damage (common-term usage, remember) is the damage that *is part of* the direct result of the failure event. In this case it is the broken rope that has suffered the direct damage. Note how the direct damage could be referred to (loosely) as 'the failure'. By now, hopefully, you should be becoming wary of using loose definitions like this. The correct term for what has happened, to the rope, is *direct damage*.

Consequential damage

For damage to be classed as consequential it has to be caused as a result of the direct damage that occurred – but it does not *include* the direct damage itself. The key is in the causal link between the two (another formal term – the *'causal link'*): one must *cause* the other. For this example the consequential damage is:

- the broken micro-switches
- the broken castings (the crane's load)
- the damage to the shop floor itself.

Note how all of these relate to components which are not the one that suffered the direct damage (the rope). Consequential damage is a particularly important concept for insurance-related investigations – mainly because it may not be one of the insured 'risks'.

Indirect damage

This is a less frequently used term which has a less robust definition. Basically, it refers to damage which results in the failed component *but* is not directly involved in the actual failure mechanism – it cannot, therefore, be considered as direct damage. It is also not caused *by* the direct damage, so it does not fall into the category of consequential damage either. Most instances of indirect damage are related to quite detailed technical descriptions of failure mechanisms – sometimes these border on the 'academic interest only' and can confuse the issue rather than clarify things. I think this term is best avoided. That does not mean that you won't see it used in failure investigation reports – you will – but bear in mind its intangibility and, frankly, its inaccuracy. You will find its definition has lots of weaknesses (as likely do some of the people who you find using it). I don't think you should take it (or them) particularly seriously. There is no 'indirect damage' in the example of the failed crane rope.

Related damage

This is another similar term which cannot be well defined, although you may see it being used. Sometimes it really means consequential damage and should be expressed as such. Steer clear of it.

Assessment

Assessment is the core of failure investigation – if you want to investigate an engineering failure, effectively, then *assessment* is what you do. The assessment activity is, unfortunately, not as discrete as shown in the framework (Fig. 2.2). It extends throughout the investigation process, over time, and includes not only the technical, engineering aspects of the failed components but also those more to do with the *context* of the investigation. A proper assessment will nearly always include some content about responsibilities (meaning liabilities) and commercial aspects (meaning money). With this in the background an assessment has to accomplish several things. It must:

- decide what happened
- decide how it happened
- conclude what the results and consequences were (and are).

This is fine, but what about that first word in the last point? Assessment is about concluding, so it must have *output* – and that output needs to be capable of initiating a conclusion. So the final point:

- The assessment stage must have conclusive *output*.

I have made this point now, early in the book, to try to demonstrate the key importance of this principle. It is probably of little use becoming involved in engineering failure investigations if you have no intention of reaching proper conclusions at the assessment stage. The assessment stage requires *precision*, it is made up of a quite tightly knit set of technical and non-technical parts, in which there is little room either for the loose definitions or bold but consensual statements that are just about acceptable in some branches of engineering. You should expect, therefore, that good assessment is difficult (though far from impossible, as hopefully we will see in Part II of this book).

The technical solution
This is shown positioned slightly off to one side of the damage/causation framework (Fig. 2.2). There is a good reason for this: a full technical solution to a failure, that is deciding some technical or design changes that will prevent recurrence of the failure, is *not always* an essential part of a failure investigation process. The main role of a technical solution is in type D investigations, where the emphasis is biased heavily towards engineering improvements, with insurance, commercial liability or other short-term financial issues firmly in second place. This doesn't mean that technical solutions don't exist for type A, B, and C failure investigations, simply that they may not always be *that* important to your client. You may find yourself slightly uncomfortable with this concept – that the technical solution is not necessarily the 'driving force' behind the activities of an engineering failure investigation. It is, however, one of the characteristics of the subject. This should help to illustrate the point that I introduced earlier, that you will need to 'step outside' the pure engineering discipline if you want to be successful at failure investigations.

The previous example of the crane rope failure provides a reasonable illustration of the role of a technical solution. The obvious solution to the rope breakage is to incorporate, in the crane's design, an overload interlock, which cuts out the winch motor when the crane attempts to lift a load exceeding its SWL. Unless, however, such a device is mandatory under the relevant legislative requirements and technical standards (and there are many different standards), then this solution is just about *irrelevant* to many types of failure investigation. There are numerous examples of type A, B, and C investigations where a perfectly satisfactory resolution of a commercial liability case or an insurance claim has been reached without any reference at all to this technical solution. It is worth emphasizing the point again:

A TECHNICAL SOLUTION TO PREVENT A FUTURE FAILURE IS OFTEN *NOT* THE MAIN ISSUE.

Reinstatement

In general usage, the term reinstatement means 'putting things right'. In the context of failure investigation the meaning is similar but a little more precise – it refers specifically to returning the failed component or equipment to the way that it *was*. This means returning the components to their pre-failure condition, whether or not this is the 'correct' engineering solution. For this reason, the act of reinstatement, although an accepted concept, particularly in insurance-related failure cases, sometimes has less-than-practical application. It is best thought of as an 'idealized' activity. In practice, even in insurance cases where failed equipment is reinstated, there will, inevitably, be changes from the pre-failure condition. Product design and detailed specifications can change surprisingly quickly, so that a component which has been in use for even a short time will likely not be exactly the same as its replacement. The situation becomes more complex when considering engineering plant and equipment, assembled from many different components – plant performance and efficiency levels are continually being improved, so it is unlikely that a piece of equipment replaced after a failure will have the same 'performance' as the old one. Most times it will be of revised design, perform better, or be made of improved materials. This scenario is common enough – almost an engineering *fact*. For this reason the concept of reinstatement must be considered in conjunction with its close relation: betterment.

Betterment

This is also an insurance-related term. It is necessary, as explained, because of the practical difficulties in reinstating engineering components to their exact pre-failure condition. Betterment is an attempt to *quantify* the differences – it attempts to ascribe a value to any changes that can be classed as an 'improvement', and can be considered as making the component or equipment 'better' for the user. Possible improvements can take several forms:

- better *performance*, such as improved flow rate, strength, productivity or efficiency
- longer *lifetime*, due to, perhaps, better resistance to wear or corrosion
- lower running costs, owing to low-maintenance design features.

You can see that the benefits of some of these features are, at least, open to interpretation and rather intangible. This is why there can be long and involved discussions about the degree of betterment that results from a reinstatement, following an engineering failure. It is not so difficult, for example, for a straightforward production machine where the output in units (nuts or bolts or whatever) per hour is clearly defined – but much more difficult for a process plant, where an apparent increase in plant efficiency could be balanced out by, perhaps, an increase in staffing or maintenance costs. The only real lowest common denominator of betterment is via the concept of net present value (NPV). Any change, as a result of the reinstatement process, that increases the NPV, can be safely classed as betterment. Note though that NPV itself can be difficult to calculate – but at least there are established techniques for doing so, which is a good start.

Settlement

You will meet the concept of settlement in most type A, B, and C failure investigations. Here is one description of it:

'SETTLEMENT' MEANS AN AGREEMENT ON COMMERCIAL MATTERS, NOT NECESSARILY ON TECHNICAL MATTERS.

True, this is the loosest of definitions, but it does have a simple accuracy about it. The common denominator of most type A, B, or C failure investigations is the question of either money or liability (which itself translates into a financial responsibility). This means that it is the concept of financial settlement which is central to 'solving' the failure investigation, rather than the need to find a unanimous technical solution. This, again, is a 'comfort zone' question – it can be difficult to accept but it is, nevertheless, a fact. You have to be careful not to misinterpret the situation, however: the definition as I have described it doesn't infer *disagreement* between the various parties on engineering matters, but neither does it require exact agreement. A broad consensus is sufficient – and probably the best thing to aim for.

A final point on settlement – you must always, *always*, focus on settlement as the target of a failure investigation. We will return to this theme several times in coming chapters. It is perhaps *the* key point of effective failure investigation.

Causation

Causation occupies the RHS of the basic damage/causation framework shown in Fig. 2.2. Deciding causation is at the centre of any failure investigation and forms one of the main inputs towards the target of settlement. There is a certain thought-pattern associated with the concept of causation – to start, forget the notion that every engineering failure has, somewhere, a single, easily definable 'cause'. This is rarely true; experience shows that failures are almost always the result of a combination of conditions and 'events' – even simple failures follow this pattern. It is best to think of causation as a family of terms, which are inter-linked, and have to be used *together* to describe the 'cause' of an engineering failure. Figure 2.3 is an attempt to show the parts in diagrammatic form – look how it is an enlargement of the RHS of Fig. 2.2. Note how Fig. 2.3 has a roughly parallel configuration, inferring that the various elements of causation do not necessarily

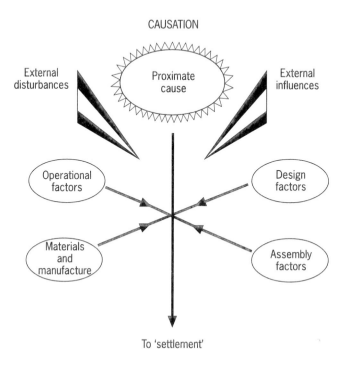

Figure 2.3   The component parts of 'causation'

have a straightforward chronological sequence. We can look at the pieces of Fig. 2.3 in turn.

Proximate cause

This is the most important definition – it is sufficiently formal, in insurance cases, to warrant a legal definition in case law:

> **Proximate cause** is 'the active, efficient cause that *sets in motion a train of events* which brings about a result' (the failure) ... 'without the intervention of any force started and working actively from a new and independent source ...': Ref: *Pawsey* v. *Scottish Union and National.*

Although phrased in a quasi-legal way, this is a very usable definition, in two ways. First, the term *proximate* comes from a root meaning 'next to' or 'nearness' – so it infers a cause that is a real and indivisible feature of a failure. Second, the definition encompasses the reality that there may be more than one attributable 'cause' of the failure, so the use of the term 'proximate cause' is a way of identifying that cause which was both dominant and effective in making the failure happen. There is even a legal maxim to back all this up: *causa proxima non remota spectatur* – meaning, broadly, that remote causes are not considered as important as more immediate (proximate) ones. Practically, there are three simple points to remember:

- 'Proximate' means *near.*
- The concept of proximate cause *accepts* that failures can be complex and that there may be more than one cause involved – proximate cause is a way to bring some clarity to the situation.
- Proximate, in this context, means *proximate in efficiency* at causing the failure, rather than proximate in time. This is inferred in the bit of the definition I have shown in italics.

Taken together, these three points should help you focus in on the proximate cause of an engineering failure. There are, of course, some weaknesses within the concept of proximate cause. The biggest one, perhaps, relates to its origin – the term was not conceived to help sort out engineering failures, rather it grew out of the wide variety of risks covered by insurance companies – so it is, at best, a lowest common denominator term. This means that the doctrine of proximate cause does not always sit easily with the technical facts of an engineering failure – you sometimes have to make simplifications (and occasional assumptions) to make it fit. The concept of a cause being proximate in *efficiency*, rather than proximate in time, can also sometimes be difficult to fit to an engineering failure, particularly one involving fatigue, as you will find the terms and conventions are not always a precise match.

Whatever the limitations of the concept of proximate cause it is fundamental to insurance-related failure investigations and used frequently for other types, particularly where there are important decisions of liability or commercial responsibility at stake. It is a good concise term. Generally, you need to guard against using loose, easily misinterpreted terms such as 'main cause', 'prime cause' and similar. These are all open to interpretation and don't mean much.

Contributing factors
These are the conditions that contribute to a failure *event* and enable the proximate cause to 'happen'. There are many possibilities:

- the design of the engineering component
- operational factors
- the environment
- wear and tear
- events; either normal or abnormal.

You can see how wide the scale can be – for this reason it is rare to find a contributing factor which exists alone, they nearly always come in multiples. Remember that they do not *cause* the failure, as such, rather they contribute towards the *conditions* that enable the failure to happen, once the cause 'appears'. There can be quite close definitions involved here – but it is important to understand the general principles, before progressing onto specific failure examples.

These, then, are the component elements of the damage/causation framework shown in Fig. 2.2. I have explained that the elements all fit together, in a broadly chronological sense, but that the framework is still only a 'model' of failure investigation – and that there will be slight differences, and changes of emphasis between the different types (A, B, C, and D) of investigation. Where does this leave us? Are we (you) now well equipped to start investigating failures – drawing conclusions, deciding causation, going for *settlement*? There are still some gaps – remember that this is meant to be a practical book, and in practice, sadly, a general framework alone (even a good framework) will not help you much. It is only the starting point. Now we have to consider the 'people' aspects – we have to look at *roles*.

KEY POINT SUMMARY: IMPORTANT CONCEPTS

1. Types of investigation

There are four main types of failure investigation:

- Type A: Commercial (including insurance-related) investigations.
- Type B: Liability investigations.
- Type C: Accident investigations.
- Type D: Research-based investigations.

They all have a similar technical structure with emphasis being placed on the *sequence of events* and categorization of the *failure mechanisms*.

2. Terminology

Terms such as *failure* and *damage* have specific technical meanings. They have to be used carefully, and in the correct context.

3. Assessment

Assessment is the core activity of failure investigation. The whole purpose of the investigation is to:

- decide *what* happened
- explain *how* it happened (causation)
- conclude what the *results* and consequences were.

4. Causation

Causation can be expressed in terms of proximate cause. This is a key definition, particularly for insurance-related failure investigations.

Chapter 3

Roles

You won't be very effective at investigating engineering failures if you don't have a feel for the various *roles* involved – and what everyone is trying to achieve.

Failure investigations have a lot of people-involvement. A typical insurance or commercial liability-related investigation can have upwards of five or six parties involved, each with various staff and technical advisers, and all with an interest in the outcome. For those in the position of providing a specialist technical input to the investigation (and that is the general standpoint I have used in this book), the situation can sometimes be confusing – you may often feel that none of the parties actually *want* to make any firm decisions, although they all seem keen for the answer to be found. This is where the technical input is important – whatever the positions of the parties (insurer, insured, litigant, etc.) involved, just about *all* the real decisions and conclusions leading up to the settlement have a *technical* basis to them. This in turn, increases the pressure on the technical advisers and engineers (you), to find conclusions which 'fit in' with the requirements of the various parties and the roles that they take.

Engineers can find this awkward – it is often difficult to balance the requirements of others with the guidelines and principles of the engineering profession. It can be a complex picture. We can start with this statement:

OBJECTIVITY IS NOT THE SAME AS NEUTRALITY.

Technical aspects of failure investigation are, as with any other facet of engineering, about *objectivity* – engineering relies on observations and conclusions being objective – but what about *neutrality*? This is not so straightforward. It's easy to start with the premise of neutrality but when faced with the structure of roles that exist in a failure investigation situation the concept of the technical investigator's input being neutral

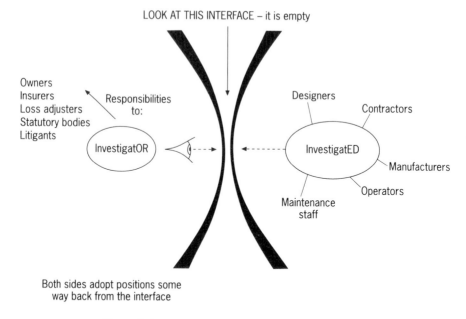

Figure 3.1   Failure investigation involves *opposition*

is a difficult one. So should a technical investigator be neutral or not? To answer this, we need to look at how the structure of roles actually works. We will look at the main role-related points. Try not to think of these as *background* to the subject of failure investigation – they are, in fact, the opposite, forming much of the detail of an investigation and explaining a little about what happens, and why, during an investigation into an engineering failure.

Investigation – opposing roles

One of the first things that you will encounter about roles in engineering failure cases is that there are two, the investiga*tor* and the investiga*ted*. You can think of these broadly as being two 'sides' of the same story. Actually this is, at best, a crude assumption – there are always at least two sides, but more likely three or four. Figure 3.1 shows the simplified case. The left side is the role of the investiga*tors*. The investigators have responsibilities to their clients – insurers, loss adjusters, statutory safety bodies – any of which may be their fee-paying clients. The purpose of the failure investigator is to 'solve' the failure for these clients. The other side of the figure represents those who are being subject *to* the investigation – the investigated. They also have responsibilities – to those parties involved in the design, manufacture and operation of the failed equipment. These are less direct responsibilities (there may be no clear contractual relationship, or fees involved) but they are responsibilities, just the same.

The two sides are in *opposition* – let there be no doubt about that – but they are not necessarily in conflict. The fixed point is that these quasi-opposing sides both have an interest in 'solving' the failure – and in doing so, attaining some kind of *settlement* (as we saw in the last chapter).

As technical adviser your role could be on either 'side' of Fig. 3.1. You may be employed to define causation with a view to deciding the validity of an insurance claim, to apportion liability or commercial responsibility, or perhaps both. All will require you to be objective but not, as we have discussed, necessarily *neutral*. We can put it like this, referring to Fig. 3.1:

INVOLVEMENT IN A FAILURE INVESTIGATION MEANS BEING ON ONE SIDE OR THE OTHER

but

YOU CANNOT BE ON THE INTERFACE

because

THERE IS NO-ONE ON THE INTERFACE – THE INTERFACE IS *EMPTY*.

Now take a close look at Fig. 3.1 – do you see how the interface is empty? This is a consequence of the roles taken in almost all failure investigations – the two sides adopt positions some way back from the interface. There are, no doubt, good reasons *why* they do this, when it would seem to make things more difficult. From the strict technical viewpoint, however, there is nothing in this confrontational approach that cannot be overcome. As long as you understand the two opposing roles of the investigators and the investigated, then you can think of them as *facilitators* in the exercise of solving the investigation rather than a barrier to it.

A closer look – offence and defence

The next step is to try to see the roles in a failure investigation in terms of *offence* and *defence*. This, again, stems from overall attempts to model and understand the situation in Fig. 3.1. Hopefully, you will find nothing that is divisive about these two terms – they mean, simply, that both roles (the investigator and the investigated) need to be offensive or defensive if they are to be effective at working towards a solution to the investigation. The key to providing the best technical input is to know *which* is required. Fortunately there is a simple guideline: a good investiga*tor* normally takes on the offensive stance, perhaps about 70 percent of the time, while the investiga*ted* (RHS of Fig. 3.1) normally relies more on *defensive* input. Note that taking on an offensive stance does not mean that you have to *cause* offence – in this context the definition is different. An offensive stance means one which is:

- active
- thorough
- questioning
- searching
- but doesn't cause offence.

The concept of offence is a little unusual – you would not be alone in finding that it sits uncomfortably with the softer attitude towards engineering problem-solving that you may be used to. It is, however, a pretty accurate description of the way that failure investigations work – and if you can get a grip on it, it can help you solve them. Without it, possibly, complex engineering failure will cause you some hardship. You might struggle – and may be better off passing the investigation to someone else.

It would be wrong to think of offence as being necessarily about saying (and feeling) *negative* things. The opposite, in fact, is truer – it is an active search for *progress* in the investigation, and you won't make much progress if all you do is think negatively. The concept of the defensive role is similar – defence does not have to mean setting out to 'spoil' the technical arguments – it is still a role that infers action, and movement. The key to either of these roles is not to let negative thinking take over – keep it out wherever you can. I have tried to illustrate the differences between offence and defence in Fig. 3.2. This refers to an (unspecified) failure problem with a gearbox. It refers to a (type A or B) commercial liability-related investigation, but the principles shown are generally applicable to all categories. Note that the statements are not specific to either the investigator or the investigated party – I have left them general – remember, though, that they refer to the same 'failure' of the gearbox.

FACT: There is something wrong with the gearbox bearings.

OFFENCE	DEFENCE
• The gearbox has failed.	• The gearbox has not failed.
• Incorrect operation can cause gearbox failures.	• We have asked the gearbox manufacturer to give an opinion.
• Gearboxes are built to published technical standards.	• All gearboxes use basically similar technical specifications.
• The gear wheel and the pinion bearing have both worn beyond their limits.	• The gearbox is fine – bearings are replaceable 'maintenance' items.
	• The gearbox is definitely not broken.
	• But if it is we did not break it.
• Look! The gearbox has failed.	• And even if we did it was an accident.
	• Anyway, bearings are easy to replace, aren't they?

Figure 3.2   Failure investigation roles – the concept of offence and defence

Professional responsibilities

Professional responsibilities are an integral feature of the landscape of failure investigations. Their importance is compounded by the existence of opposing positions in the investigation. Granted, opposing positions can place a strain on the proper execution of professional responsibilities but they can also help by throwing them into *relief* – making them clearer to understand. There is nothing particularly new about the professional engineering responsibilities involved in a failure investigation situation, but it helps if you can keep them in focus. There is nothing altruistic about this – it will help *your* situation. Figure 3.3 is an attempt to show how professional engineering responsibilities mesh with

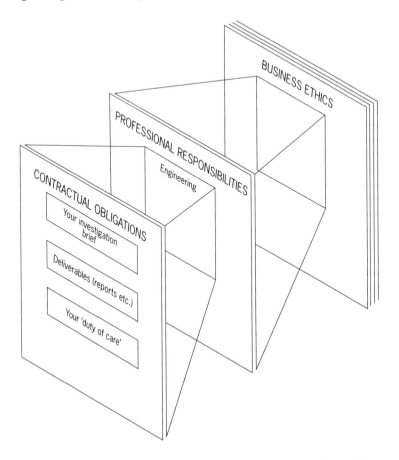

Figure 3.3   Your professional responsibilities – how they fit together

those of your direct contractual obligations (to your client) and with the wider envelope of professional and business ethics. Acting responsibly – with integrity and, of course, honesty – is an important constituent of all three responsibility 'levels' as shown. So much for *your* responsibilities representing either the investigator or investigated – but what about the other parties, those involved directly in the results of the failure investigation? Roles and responsibilities are all important here – the way that you draw conclusions about engineering failures needs to be 'in step' with the responsibilities of the other parties. Crudely, this means that your technically based report will be used to decide whether someone has done their job, as expected of them, or not. Two subjects take priority here: duty of care and design responsibility.

Duty of care

The idea of 'duty of care' is inherent in engineering contracts, almost without exception. I am sure that there are specific legal definitions of it but the plain-language definition is better:

Duty of care means being *responsible in* what you do.

This is a little different from being responsible *for* what you do, which infers liability and the spectre of *negligence* – an altogether different thing, with dangerous legal meaning. All parties involved in designing, manufacturing and using an engineering product have this duty of care. It is a common-law requirement and applies, therefore, whether or not it is mentioned in 'the contract' between involved parties. Its role in failure investigations can be decisive in deciding liability – a party that has not compiled with its duty of care could be exposed to strong arguments (the offence) that it contributed significantly to the failure. Equally, proven compliance with all duty-of-care requirements can form the basis of a case for avoiding liability for a failure (the defence) – although it is rarely *all* of it.

Design responsibility

The question of *design responsibility* often forms part of the offensive or defensive activities of the various parties in a failure investigation. It is related specifically to the engineering aspects – you will often hear engineering component failures blamed initially on 'bad design'.

The problem with the concept of design responsibility is that it is fragmented into several parts. Take the simple example used earlier, that of an engine or turbine-driven gearbox (the exact type is not important). No single person or company actually designs, themselves, all of the engineering components that comprise the operating gearbox. There are at least four 'levels' of design to consider, each contributing to the task of providing a gearbox that works. Figure 3.4 shows one way to look at the breakdown of this responsibility – note the four 'levels' of design responsibility: technology; process system; equipment item; and equipment component. There has been design input to all of these levels, starting with the overall 'technology' design (in this case the technology of helical gear trains), down to the specialist design input needed to decide the correct sizes, loadings and materials of construction for the bearings and other individual components. From this you can see that the responsibility for a design *fault* (if there is one) is not easy to define – the origin of the fault may be at any of the four levels or, more likely, at more than one level. There is often no easy answer to this. Fortunately,

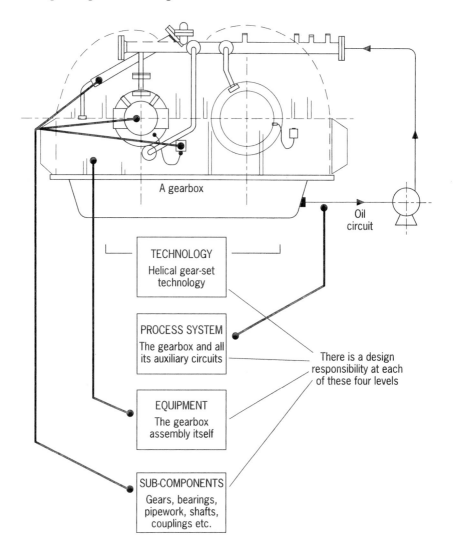

Figure 3.4   The breakdown of design responsibility

there is one fixed point – each design level has, as its backbone, the requirement of 'duty of care' outlined earlier. In practical failure cases, the role of the investigator frequently involves looking for (technical) evidence that duty of care has been fulfilled. The important message, however, is to understand the complexity of the design process and to remember that, when the question is orientated towards design responsibility, there can be several levels of design that need checking.

The role of money

Do you think that this statement is true?

ENGINEERING FAILURE INVESTIGATIONS ARE ABOUT MONEY.

There is certainly some evidence for it – we saw in Chapter 2 that, of the four main types of failure investigation, money is the driving force in at least two of them, and has a part to play in the others too. From a pure technical viewpoint however, the issue of money is secondary – it is the engineering components that are under scrutiny during a failure investigation, not the financial affairs of the company that made them. One of the overt purposes of this book is to help you draw the link between the pure technical parts of a failure investigation and the other, broader, issues that surround them. They *all* form part of the picture, and so cannot be ignored, but do need to be seen in the right context.

Commercial considerations are an inferred part of the brief (your brief) of providing professional technical input to a failure investigation. For commerce, read *money*. You cannot divorce it from the results of your investigation and you have a duty of care, to your client, to understand the financial implications of your own technical conclusions. This is part of your professional responsibility. It is best if you can see money as being *linked to* the failure investigation, in several places, rather than being an integral part of it. What it is not wise to do is to let the possible financial consequences of your technical conclusions *lead* the failure investigation. This is a classic weak link, and rarely results in good engineering answers – nor, surprisingly, does it often lead to a clear, quick settlement, as maybe you would think. Keep the financial issues in context – just don't forget that they are there.

Another issue is the role that money plays in the 'settlement' arrangements. This is mainly of concern in insurance-related failure investigations. The technical issues of a failure case are normally a precursor to any settlement 'bargain', and so they have a part to play, but there is a point in settlement negotiations beyond which the technical and engineering issues tend to recede. It is not easy to anticipate when this will occur – it is unique to each individual investigation – so you may have to accept a less-than-perfect understanding in this area, and develop a certain intuition for when this transition occurs.

A final word of advice – don't get too emotive about the financial issues that lie behind the results of failure investigations – money is only a *measure* (albeit a good one) of value, it does not confer value in itself. Don't let it cloud your engineering judgement.

Victory and defeat

The final point on roles in failure investigations is the question of whether, after all technical conclusions have been decided and settlement reached, there is a winner and a loser. Most types of negotiation procedure, which are competitions, of sorts, end with a solution that is a combination of winning and losing 'elements', but with areas of general mutual benefit to both parties. So there is a 'win-win' component as well. Engineering failure investigations are similar, the only significant difference being that the financial negotiations are usually retrospective, i.e. about costs that have already been recovered, and are, therefore, predetermined. There are many areas of engineering contract negotiation that are like this, so the idea is hardly new.

Figure 3.5 is an attempt to illustrate the situation (you could superimpose this over Fig. 3.1) regarding victory and defeat of the opposing

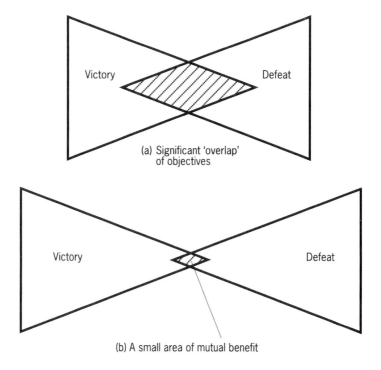

Figure 3.5   Settlement – victory and defeat, or mutual benefit?

sides. It shows the result of the final *settlement* of the investigation. Note the size of the shaded 'overlapping' area in the centre. This represents the area where any settlement is *genuinely* of benefit to both opposing sides – agreements on such things as the extent of damage (for insurance-related cases) and the broad nature of the failure often fall into this 'mutual interest' area. The question is: How big is this overlapping area likely to be? Does most of the settlement end up as 'win or lose', or is it generally biased towards mutual benefit? The answer, of course, depends on the individual circumstances of the failure and the type of investigation. In terms of *advice*, the best recommendation I can offer is that you visualize the situation as being like Fig. 3.5(b), rather than Fig. 3.5(a), with a small 'overlapping' area as shown. This will help your focus. It doesn't mean that the particular failure investigation in which you are involved will necessarily *be* like this, but it is a good model from which to start. Don't worry too much about the accuracy of this with respect to your technical role – the bigger picture is, anyway, more the domain of your clients, rather than their technical advisers.

KEY POINT SUMMARY: ROLES

1. Understanding the roles

It is important to understand the roles that the various parties play in a failure investigation. There are often many other (non-engineering) people involved.

2. Opposition

A failure investigation generally involves several parties who are in *opposition*. A technical investigator is in the difficult position of having to be *objective* (but not necessarily *neutral*).

3. Professional responsibility

Investigators have a professional responsibility and a 'duty of care'. This means that you have to be responsible in (and for) what you do.

4. The role of money

Commercial considerations are nearly always an important part of failure investigations. There is absolutely no need to be frightened by this.

5. Victory and defeat

All investigations have to end in agreement eventually. Sometimes there is a winner and a loser and sometimes all parties are happy with the technical and commercial results.

Chapter 4

Strategy and tactics

Engineering failure investigations, we have decided, involve victory and defeat. This means that, somewhere along the line, there are going to be matters of discussion and interpretation. In short, there is going to be *disagreement*. Some of this will involve (and affect) the work of technical advisers – it would be rare indeed to find a failure investigation in which all the engineering advisers and their principals disagree only on financial matters. To help you deal with this, you need to be organized in what you are doing – you cannot, frankly, hope to be effective in failure investigations unless you have:

- An appreciation of the *strategy* behind what you are doing.
- Access to a set of simple *tactics*.

No doubt it is possible to take part in engineering failure investigations, and to be a technical adviser to one or more of the parties, without these two things. I am sure that engineers do, all the time. This is fine, but the question is: How effective do you *want to be*? While the technical skills involved in understanding and analysing engineering failures are quite traditional and straightforward, the ways in which the process is conducted, and the methods by which conclusions are expressed (and used), are not. There are some well-rehearsed techniques and practices which have developed and which are specific to the generic discipline of failure investigation. These can save you a lot of time and effort – and will mean that you will be more effective in your role. Your clients might even 'win' a higher percentage of their cases, because of your input.

Sadly, there is no overall 'secret' to successful failure investigation (I wish there were), but understanding the general strategy, and a few tactics, is a good place to start. Strategy, basically, is the overall plan of a failure investigation and the path that it takes towards settlement. It has to accept the problems and real pressures that exist when a failure occurs. Tactics are more specific – they are discrete elements of practical advice that you need in order to *participate* in the investigation. You can think of tactics as being a set of basic skills which you bring to a failure investigation – some are to do with engineering but some are wider, related to subjects like information gathering, reporting and making decisions. We will look at strategy first; and then at some tactics.

Strategy

The overall strategy of a failure investigation is, in some ways, set by the underlying framework of damage and causation that we saw in Fig. 2.2. The definitions of the various types of event and causation also help. There is little that you, as technical adviser, can do to change this framework, the best you can hope for is to try to fit it in with the way that the investigation can be expected to progress. The first point in your strategy should be: *get organized.* There are several elements to this:

- Plan for the *course of events.* We have already discussed the fact that the parts of a failure investigation follow on from each other in a more or less predictable way (Figs 2.2 and 2.3). Try and plan how you intend to approach each stage – so that you won't end up thinking (and even worse, talking), without adequate preparation.
- Anticipate the *roles* that everyone will adopt. This was introduced in Chapter 3 and we will look at a case-specific example in Part II of this book. A little forethought about the roles of the various parties involved in a failure investigation undoubtedly pays dividends – you won't waste so much time and effort. Pay particular attention to the role of *your clients* – they will benefit from receiving your technical arguments in a 'way which fits' with their role, and what they are trying to prove, or disprove.
- Try for effectiveness, but also for *economy.* Make your investigations and draw your conclusions in the most economical way you can – there is little real advantage to be gained by 'drawing out' an investigation, or embarking on continuing searches for more information or further research. This annoys everyone. Do your investigation carefully, applying the idea of offence (or defence) as set out in Chapter 3, and then draw your conclusions. *Finish* it. You will benefit in the longer term from using this type of approach – it will increase your credibility as a failure investigator.

Strategy – understanding 'focus'
It is easy to see failure investigations purely from an engineering viewpoint – as a neat technical puzzle with an equally precise set of

answers, just waiting to be discovered. This is fine. It is the best focus for technical advisers to take. The overall picture, however, is not quite so straightforward, so part of a good strategic approach is having an appreciation of some of the different foci that exist so that you will recognize them when they appear. This will reduce the risk that you will slip away from the all-important engineering focus, perhaps into a different one, by mistake.

Figure 4.1 shows a typical focus 'set' for a failure investigation – in this case for an insurance-related case. The engineers' focus is shown, purely for convenience, at the centre of the figure. Any insurance claim investigation, even a simple one, will possess the other foci as shown. Note that the further you radiate outwards from the central engineering focus, the *less* the technical knowledge (and subsequent technical competence) of the focus covered. This is merely a way of illustrating that the insurer of a complex piece of equipment will have a different

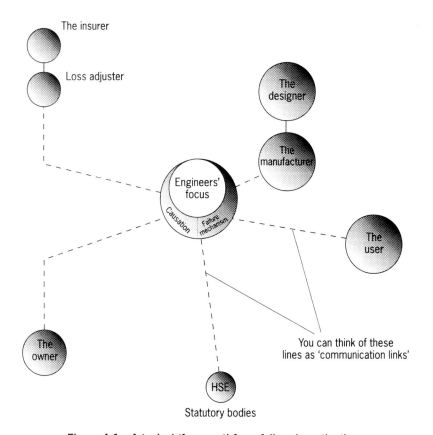

Figure 4.1   A typical 'focus set' for a failure investigation

level of knowledge of the technicalities of that equipment to that of the loss adjusters. They, in turn, generally know less than their appointed technical advisers. Perhaps this is really only stating the obvious – but you would be surprised at the frequency with which advisers' technical reports, particularly their conclusions, do not take this into account. It is as if it sometimes comes as a surprise.

There are two further important points to note from Fig. 4.1. The first is the content of the engineer's focus (the central 'ball'). The core content of this focus is the *twin levels of conclusion* – the separation of the description of the failure mechanism from the definition of its cause. The message is that conclusions cannot be best described in a single 'level'. The other important point is the existence of the lines linking the engineers' focus with the others – these represent lines of *technical communication*, about the extent of damage events, causation, and the other concepts discussed in Chapter 2 that make up the overall territory of a failure investigation. It is sometimes necessary to *use* these communication links, but not so often that you lose your engineers' focus on the failure. You can think of the communication links as being part of your focus. One of the most popular strategic mistakes is concentrating only on the pure technical view of a failure, without appreciating that there are other equally valid (but different) viewpoints. If you make this mistake, the end result will be that your technical conclusions on the failure mechanism and causation will be poor. They will not be well-accepted by the various parties concerned.

These points represent the backbone of a good, strategic approach to failure investigation. There is much more that could be written on the subject but there is a danger in going too far – it is easy to cloud the main issues. Too much strategy can be self-defeating, the common 'paralysis by analysis' syndrome can become a sad feature of failure investigations, if you go too far.

Tactics

The procedure of engineering failure investigations can vary between different types of investigation but we are fortunate in that some of the more basic *tactical* elements are common to them all. Tactics are those discrete areas of practical advice which can help you participate effectively in failure investigations. Each of these areas is only a small part of the picture but, taken together, they can help bring organization and structure to what you do. There is a set of basic tactics relevant to engineering failure investigations, complemented by a number of

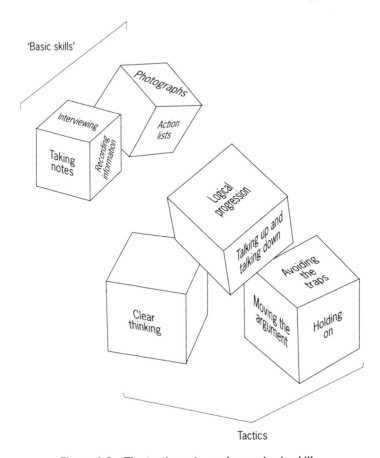

'Basic skills'

Photographs

Interviewing

Action lists

Taking notes

Recording information

Logical progression

Talking up and talking down

Avoiding the traps

Clear thinking

Moving the argument

Holding on

Tactics

Figure 4.2   The tactics set – and some basic skills

activities which are best described as 'basic skills'. Figure 4.2 shows them in diagrammatic form – we can look at them in turn.

Clear thinking

Clear thinking is undoubtedly a tactic rather than part of strategy. Clear thinking means not drawing conclusions based on poor factual inform-ation, and taking care to move step-by-step through a problem rather than travelling in mental circles, or becoming confused. In the context of engineering failure investigation there are several areas of clear thinking which are of particular interest:

- *Getting rid of preconceptions.* It is important not to harbour too many preconceptions about the way that engineering components *might* fail – causes of failure are fairly predictable but it is still dangerous to expect too much repetition of specific failure mechanisms and causes that you have seen before. This doesn't mean that you can afford to ignore experience, merely that you should leave room for some judgement in each individual failure case. You will be surprised at the number of times you will find 'new' events and causes just when you thought you had seen them all before.

● *Listen to first impressions.* There are often lots of first impressions around as to the 'cause' of an engineering failure – the larger and more serious the failure, the greater the number of opinions you will hear. It is always good practice to listen to these first impressions – don't be frightened about forming some of your own also. Despite some viewpoints to the contrary, experience shows that they are often very useful pointers towards finding the causes of failure (both proximate and contributory). First impressions can contain a surprising amount of engineering intuition and judgement – in some people more than others. Note that I said *listen* to others' first impressions – but be careful not to comment, or talk loosely about them, until your views have become better formed.

● *Learn to identify causality.* This is not quite the same context as the idea of causation introduced in Chapter 2. The key skill here is in identifying causal *relationships*. This is a form of logical thinking, best illustrated by an example: Here are two statements:

   (a) The gearbox oil supply was interrupted, causing the bearings to fail.

   (b) The gearbox was not designed for 'overload', so it failed.

The first statement (a) has a clear causal relationship – the bearing failure can be clearly linked, technically, to the event of oil supply failure. You could collect conclusive technical evidence to show how a lack of oil would cause the bearing to fail. The statement is absolutely causal – the oil interruption *caused* the failure. The second statement (b), however, is clearly not causal. Whereas the two parts of the statement – 'The gearbox was not designed for overload' and 'it failed' – may both in themselves be accurate statements of fact, there is nothing that undeniably links them together. One does not necessarily cause the other. This difference between causal and non-causal statements is a significant factor in analysing and reporting failures. Causal statements encourage (and demonstrate) clear thinking while non-causal statements are more likely to be linked to thinking which is muddled – it will have flaws. Technical conclusions, and reports, should rely on causal statements to be effective.

The secret of clear thinking is practice. The more failure investigations you do the clearer your thinking will become; you can help the process by becoming familiar with the three areas I have introduced. They are certainly not a panacea, but they can help you move in the right direction.

Logical progression

This follows on naturally from clear thinking, but the main emphasis is on the *sequence* in which technical issues are addressed. It applies whether you are organizing your approach before investigating a failure, or constructing your report and conclusions at the end of it. The rule is simple:

Use a logical *technical* progression.

This means that each step of your investigation activity (or your report) should follow on from the last in a way that reflects the chronological procedure of design, manufacture, and the steps through to failure, of the engineering component in question. For manufactured components (and it is difficult to think of any other type), this means following the five main procedural steps, in order, as follows:

- design
- manufacture
- assembly
- operation, i.e. use
- failure.

Figure 4.3 shows an example of the use of such logical progression in the structure of a failure report. This is a simple, abstracted example from the case of a failed steel bridge deck. The bridge deck is composed of longitudinal steel I-section beams (*members*) fabricated from 20 mm steel plate. One of the longitudinal members has cracked, causing the bridge deck to sag and become unsafe. Note how the text progresses through the five steps above. I have also shown an example of poor logical progression – look how it is confusing, and lacks precision. Please try to grasp this concept of logical progression – it will help you become more effective at failure investigation.

Talking up and talking down

Effective failure investigation sometimes relies quite heavily on this tactic. I did not invent it; you will find it in frequent use in all categories of failure investigation except, perhaps, in pure research (type D)

Abstract (a): good logical progression

The longitudinal member is designed to the structural steelwork Standard BS 5400, incorporating fabricated I-beams with a factor of safety exceeding 7.	*Design*
The member is fabricated using approved submerged arc welding (SAW) techniques.	*Manufacture*
Web/flange alignment was measured and found to be within specified limits.	*Assembly*
The loading of the member was checked and found to be as shown in the design drawings. Maximum shear force loadings did not exceed specified limits.	*Operation*
The member failed, following propagation of a crack from the toe of an over-convex weld near one of the strut positions.	*Failure*

Abstract (b): poor (awful) logical progression

The loading of the member was found to be as shown in the design drawings.	*Operation*

so

This could not have caused the failure which was seen on the weld strut.	*Failure*
The weld alignment was acceptable.	*Assembly*

and

BS 5400 covers this in some detail.	*Design*

also

The flange alignments were acceptable, within the design limit standard.	*Assembly (again)* *Design (again)*
It is possible that the failure was caused by the method of manufacture.	*Manufacture*

Figure 4.3   Examples of good and bad logical progression

examples, although even there it is used, albeit in a slightly different form. Talking 'up' or talking 'down' involves placing various levels of emphasis on a technical fact, to make it seem more (or less) important, respectively. Note that it is not the technical fact itself that is in dispute, it is its importance to, for instance, deciding the category of a failure or its causation. The main use of this tactic is in agreeing the relative importance of the various chronological stages by which a component has failed, rather than to agree the extent of damage (which is normally fairly obvious). We will see later that 'the failure' of an engineering component can, in purely technical terms, be divided into four stages: initiation, damage, fracture, and rupture (break), and the contribution of each of these stages to the overall 'failure' is always an area of lively discussion – so you can expect to see talking 'up' and talking 'down' playing its part.

One short word of caution – make sure, if you use this tactic, that you use it only for the purposes of *expression*, not misrepresentation. It is not good practice to try to misrepresent technical facts. The main reason for this is practical rather than moral; you will always find people with better technical ability than yourself in specialist areas and some of them can really make you struggle. At best, you will lose your credibility. So, talk up or talk down, by all means, but don't misrepresent technical facts.

Moving the argument

This is, unashamedly, a tactic for *defence*. While 'talking up or down' was to do with the amount of emphasis placed on a technical fact, this one is about moving *between* technical facts; from one to another. Don't confuse this with any aspect of the financial negotiation that may be part of the settlement activity of a commercial or insurance-related failure – we are talking about purely technical disciplines here. The purpose of 'moving the argument' from one technical fact to another is to encourage everyone to reach conclusions – a lot of people can waste a lot of time struggling with the consequences of a particularly difficult technical description, while there is often a much easier, technical way to express it. For example:

- *Static loadings* are easier to explain than complex fatigue or creep effects.
- Plant *operation* can be discussed more objectively than can plant *design* (which is multidisciplinary).
- *Total failure* 'events' generally have clearer failure mechanisms than do instances of partial failure (which is difficult to define, anyway).

Expect to find instances where you *have* to 'move' the technical argument, as in the examples above, in order to make progress in a failure investigation. While I have said that this tactic is a defensive one, you should guard against the temptation to use it for evasion, i.e. to try to 'change the subject' when a technical fact is not in your favour. This is not its purpose – it is there to improve the situation, to encourage conclusions and decisions about causation, not to help block them.

Holding on

Failure investigations, because of their connections with liability and blame (and money) can become an emotive business. *Not everyone will like what you conclude.* You can reasonably expect three things:

- differences of technical opinion
- technical criticism
- other types of criticism.

How do you reach clear technical conclusions through all of this? Technical knowledge is all-important, as of course is experience (this book might even help you in a few areas), but you will still find your technical conclusions, particularly those about causation, coming under pressure. So 'holding on' means having *confidence* in your own technical ability and its conclusions. I cannot say much more about it – only that it is an important tactic in the failure investigation process.

Avoiding the traps

It is good tactics not to fall into the various traps which are always present in the technical side of failure investigations. The three main traps are common 'destinations' for engineers working in failure investigation so it is best to anticipate, and avoid, them if you can. They won't destroy your chances of making a successful failure investigation – but they can certainly make things more difficult.

Circular arguments

Circular arguments abound in some engineering failure investigations. They occur most frequently when discussing the role of the *design* of a component in causing, or contributing to, a failure. The reason for this is that the design process itself is rather circular and iterative, so it is not always easy to make a single definitive statement of causation without it being heavily qualified by reference to other aspects of the design. Figure 4.4 should explain this more clearly – it is based on one of the design 'loops' inherent in the subject of single-helical gearbox design and shows how the parameters of gear size, thrust bearing size and

The design process for a single-helical gearbox

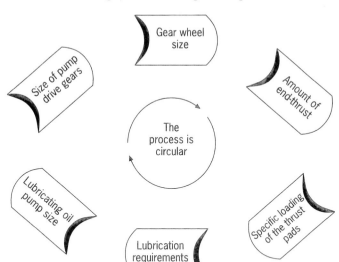

Figure 4.4   A typical design 'loop' – and a possible circular argument

lubricating oil (LO) flow, are nested and interrelated. The process of design involves moving clockwise around the design loop – several iterations may be needed before a design is finalized. You can see the problem with trying to identify a single design element which may have caused the failure – any of the elements shown could be said to carry a part of the blame. This is an example of a circular argument.

So how should you treat circular arguments? It is of little use pretending they don't exist – one way to deal with them is to consciously *break them up* into smaller pieces so that, at least, when you report about what *caused* a failure, you can make the situation look clearer than it actually is. This is still difficult – you have not changed the circular form of the design process just by being selective about what you report. A better way is not to get too involved with the detailed design of failed components – you can *comment*, perhaps, about design features but it is often difficult to justify a categoric statement that the design of a failed component was 'wrong'. There are always good (but equally circular) technical arguments that can justifiably defeat you if allegations are made about others' designs. So:

ARGUMENTS ABOUT DESIGN CAN BE CIRCULAR – BE PREPARED FOR THIS.

Never-ending research work

This is a very easy scenario to fall into – and, paradoxically, it is often the better-experienced, more thorough engineers that drift down this path in failure investigations. The idea, which is an errant one, is that the more research work carried out on a failure investigation, the more conclusive and incisive will be the results and decisions on the failure mechanism, and its causation. I have never found this to be true. In most varieties of commercial or insurance-based investigations (i.e. types A, B, and C as introduced in Chapter 2), the amount of additional 'research' work needed to enable robust decisions (and *settlement*) to be reached is very small. This is because, at the technological level of commercial failure analysis, most of the relevant research work has already been done, and recorded, somewhere. Do you think that your case is the first crane, gearbox, bridge, or whatever, that has failed? This is where your personal information-research skills come in – the relevant data on creep, fatigue life, corrosion rates etc. are not always easy to find but will be available somewhere, if you look. This is a quicker and more reliable way of getting data than by commissioning your own research work.

My advice is to keep the number of 'research and test' activities of a failure investigation as small as possible. For tests that are essential to help diagnosis of the failure mechanism, make sure that the limits and objectives of the tests are clearly set before the tests start. Simple metallurgical tests fall into this category – they can be helpful in providing 'pointers' towards causation but only if the scope of the tests is carefully chosen and then properly controlled. The best way to commission metallurgical tests is to have a hypothesis to work from (e.g. the crack started from a surface defect) and then let the test prove or disprove this assumption. This way you will get a useful answer. It is rather pointless just to ask for 'metallurgical tests' on a failed component. There is a less-than-even chance that the conclusions will contain anything useful and you will be at risk of being drawn into a 'further research' programme, with all the attendant dangers I have described. So the key tactical points are:

AVOID ONGOING RESEARCH WORK.

And, for those simple tests that are required:

USE A HYPOTHESIS APPROACH TO GIVE SOME DIRECTION TO THE TESTS.

'Groupthink'

This is an interesting one. 'Groupthink' is what happens when a group of people representing different, and often quite diverse, parties all start to think alike. Suddenly, a particular technical answer or solution (normally the lowest common denominator one possible) becomes attractive to all. Previous sceptics and dissenters miraculously become converts and everyone in the group moves to convince themselves (and each other) that this conclusion, 'just has to be correct' and 'how could there possibly be another answer?' This is groupthink, sometimes referred to, if a little unfairly, as 'nodding donkey syndrome'.

Groupthink occurs for a variety of reasons. Very complex failure investigations sometimes have it – everyone gets baffled by the complexity, and the technical detail, to the point where they all desire a simple answer, even if it is not the best one. The other factor that encourages it is *timescale* – long drawn-out investigations with too much technical research (we know all about that) can become tiresome for the participants, leading as before to the unanimous search for a consensus solution, however facile. There is excellent correlation between these two types of investigation and the appearance of 'groupthink' conclusions on both the categorization of failure mechanisms and their causation – a long timescale, add a little technical complexity, and out come the donkeys. Groupthink seems to be mainly a technical phenomenon and doesn't extend to the *settlement* part of failure investigation. This is part of the problem – if 'groupthink' actually led to easy settlement of insurance- and liability-related failures then its existence would be good news for all concerned. In practice, however, the high level of apparent technical consensus that may be reached rarely extends to the financial and liability implications of the failure – so you won't see an 'upside' in the form of streamlined settlement of the commercial affairs.

So, how should you deal with groupthink – what is the *tactic*? The best way is to allow it to take its course of events – let it roll. This way you won't be accused of holding up the investigation. What you should *not* do, however, is to let this illusion of consensus become 'the answer' to the failure investigation. You have to guard against being seduced into thinking that it will produce proper engineering decisions and conclusions on events, failure mechanisms and causation. Once the 'groupthink solution' has receded (as it inevitably will, when it has to be related to the settlement stage), then *you* will still be left with the engineering failure to solve. You may have to start again – progressing through the concepts put forward in Chapter 2 in a logical and structured way, thinking about offence, and defence, so you can start leading towards some proper technical decisions. Can you see how groupthink is a *negative* feature of failure investigation? Try to formulate a sound tactical approach to it, so it won't cause you wasted effort.

This concludes the outline of the basic tactical 'set' shown in Fig. 4.2. You should find them appearing regularly in engineering failure investigations – you can find some gentle examples in Part II of this book. It is important to gain an understanding of the basic tactics of failure investigation – the process is not, as it might appear, just a rather random series of technical events and opinions, there is a particular structure to it. Formal tactical considerations apart, there are a number of basic skills that you need to help your work on failure investigation (the upper part of Fig. 4.2). These join together with the 'tactics set' to form an organized approach.

Basic skills

The basic skills that you need are mainly to do with techniques of *information gathering*. I have excluded consideration of your engineering knowledge, and of the skills of reporting – both are covered later in the book. Information gathering is a core activity in failure investigation. Several *principles* of information recording are also important – raw information in itself will not necessarily be useful to the investigator, unless it complies with these important principles:

- Information needs to be *relevant*. This means that it must be directly related to the specific failure and its investigation, rather than the design, manufacture or operation of similar types of components, or to engineering practice in general. Perhaps this sounds obvious, but you would be surprised at the number of times you will find 'generic type' information being presented as evidence for causation in a failure investigation. There is a certain amount of clear thinking required when collecting information. Figure 4.5 demonstrates some typical bits of information, of varying degrees of relevance to the failure described – it relates to the broken crane rope mentioned earlier in the book. Note how the statements in the right-hand column, although apparently more detailed and comprehensive than those in the middle column, suffer from being too general. This is not to say that the points are not *true*, just that their lack of precision reduces significantly their relevance to the investigation procedure. They have no *edge*.
- Information needs *logical progression*. I have already covered the

The context	Relevant information	Not-so-relevant information
1. The crane rope has broken.	The rope had a factor of safety of six.	Crane ropes will break if overloaded *(statement of the obvious)*.
2. The crane rope broke due to overload.	The overload was 130 percent of SWL.	The rope was proof-tested to 125 percent SWL *(a true statement of fact but not directly relevant to this failure)*.
3. The rope failure occurred due to a dynamic loading.	The dynamic loading was caused by 'jerking' the load from the floor.	Crane ropes should be designed to accommodate dynamic loadings *(this is opinion, not a fact about the failure)*.
4. The crane was being operated in 'manual control' mode when the rope failure occurred.	Operator X was operating the crane when the failure occurred.	Manual control mode allows some automatic protection devices to be overridden. *(So, were they overridden or not?)*.

Figure 4.5   The principle of relevant information

general principles of logical progression in Fig. 4.3. As a general requirement all information presented in failure investigation cases needs to be structured in a way which makes it easy to understand, technically. For instance:

- When describing a piece of equipment, start with an overall description, and work down to the smallest relevant component. If you do it the other way round, it will be confusing.
- If you *have* to make 'generic' engineering statements, do them first – get them out of the way before you discuss specific events or causes of a particular failure.
- As in Fig. 4.3, address and report on issues in their logical, chronological series: design, manufacture, assembly, operation, then failure (and also reinstatement if this is an issue). You will be surprised in just how many situations you can use this technique – you don't have to use all these headings, but try to keep what you use in their proper chronological sequence.

● Make information *conclusive*. Inconclusive information is one of the biggest areas of mischief in failure investigations. Sadly, it is a common occurrence. Presenting inconclusive information, in any form, will only delay the proper progress of a failure investigation – it will have to be reappraised and reinterpreted, possibly by others, and the investigation will start to wander. It will lose some of its technical challenge, and *freshness*. You may be partly to blame – keep information short, blunt if you have to, make it *drive* a conclusion, and you won't have these types of problems.

Taking notes

Taking notes is one of the core technical skills of failure investigation – however limited your involvement, you will find yourself having to take notes at some time or other. The content and format of these notes are *important* – notes taken 'on the spot' during failure cases can have the status of testimonial judicial evidence and therefore can be called for disclosure in court, if a case progresses to litigation. There is a big difference between a set of notes that records your findings effectively and those that are merely 'jottings'. While jottings may make perfect sense (to you) when you write them it is unlikely that they will do so in a few months time, when the fine detail of what you saw has become a little clouded. Another problem with jottings is that they are difficult, if not impossible, for other people to understand – you will find them difficult to explain, and so people may start to doubt their (and your) credibility.

To be credible, notes need to be precise and to be assembled to some kind of structure. 'Precision' means that the statements need to be accurate reflections of what you found *and* they need to be self-supporting – to tell a little story in themselves. This means that one- or two-word 'jottings' will rarely suffice – it is necessary to add supporting words to make the notes clear. Try to think of this *before* you write your first note and the procedure should follow quite naturally – it is not difficult. The second point of the 'precision' requirement is to make clear what is fact and what is *opinion*. There is nothing wrong with expressing opinion in your note-taking as long as it cannot easily be confused with fact (others can disagree with opinions, but not so easily with facts). Figure 4.6 shows a straightforward failure case – a broken set screw. This is a simple failure 'event' caused broadly by bending fatigue; the exact failure mechanism is not important for the example. The figure shows two sets of notes taken, at the same time, during the preliminary inspection of the failed set screw. Note how the lower example, Fig. 4.6(b), consists of 'jottings' rather than proper notes – the terminology is loose, the statements are mixed-up and the technical accuracy is poor. There is no *precision*. The upper example,

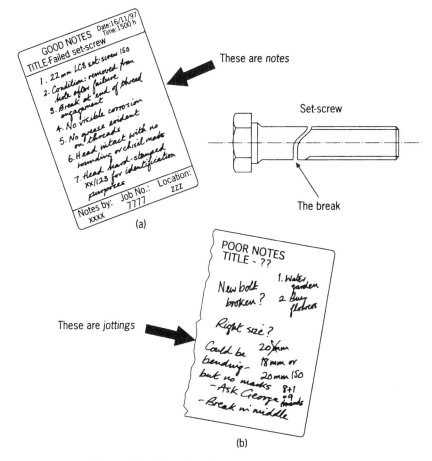

Figure 4.6   Example of good and poor note taking

Fig. 4.6(a), is better – the statements are more self-supporting and are arranged in a logical numbered order.

The way that you *structure* notes also influences their effectiveness. Figure 4.6(a) is built up along reasonably logical lines, but with the disadvantage that the logic is more apparent to the note-taker than the reader (could you tell, for instance that I was thinking more of opinion than fact when I wrote points 3 and 4 of these notes?). One way to improve the obvious logic of your notes is to use *notemaps*. Notemaps are a way of setting out notes in a diagrammatic form – this lets you show the basic logic and chronological relationships between the points. You can also separate facts and opinions, spatially, so there is less chance of confusion. Figure 4.7 shows the salient points of Fig. 4.6(a)

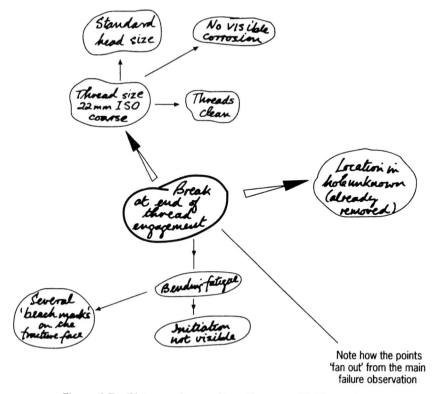

Note how the points
'fan out' from the main
failure observation

Figure 4.7   'Notemaps' – an alternative way of taking notes

presented in 'notemap' format – note how the various points flow
'outwards' from the record of the main failure observation. This is a
simple example – notemaps are most useful when the number of
observations gets above about 25 points: it is much easier to see the
relationship between them when they are expressed graphically rather
than as a long vertical list of partially-interconnected statements.

Recording information
Note-taking is only one way of recording technical information at the
'point of investigation'. There are several others which you may see
being used:

● *Letters.* There is nothing inherently wrong in recording technical
  information in the form of letters sent between the various parties
  involved in a failure investigation. The weakness, however, is the
  limited technical coverage that is possible in a letter – there is rarely
  room for the detailed technical justification that is possible in a

report. The best advice is to only mention in letters those technical aspects that you have covered elsewhere, preferably in a technical report. Let the letter repeat, or refer to, what is in the report, but don't introduce *new* technical information in it. Despite their popularity, letters (and memos) are really not very good vehicles for communicating technical detail. If you do write letters, follow the two often-repeated guidelines:

- Use separate *paragraphs*, each containing one, and only one, technical 'thought'. I have hundreds of sample letters where paragraphs are horrendously long, causing the relevant technical facts to be obscured by a wall of text.
- Use a *bold typeface* to highlight key words and phrases. Stick to a maximum of three or four highlighted items per letter, or they will tend to merge together.

- *Reports.* Report content will be covered in detail in Chapter 9 of this book. For the moment, I will only say that there is little room in failure investigation for loose or inaccurate reporting. Failure reporting needs to have a formality to it, perhaps surpassing that acceptable in other disciplines of engineering. Reports can have legal standing – so treat them seriously. You will benefit from this type of approach.

- *Sketches.* These are a useful and often undervalued method of recording technical information in failure investigations. They can have the same legal status as written notes and so benefit from the same attention to accuracy and precision. All your sketches should be annotated, with careful notes made about size, scale and other practical engineering aspects that may not be obvious to someone looking at them for the first time. Sketches used for recording *observations* made about failed components need to be clearly marked as such, so that they cannot be confused with normal working drawings of the component. This is an important distinction.

- *Minutes of meetings.* These have the dubious distinction of being just about the *worst* way to record and transmit technical information about engineering failures. There are far too many political points and personal viewpoints bound up in most sets of minutes to make them suitable for objective technical observation. Avoid them at all costs during the earlier 'technical' stages of an engineering failure investigation. You might be forced to accept their usage during the later 'settlement' stages, but bear in mind their dangerous weak points.

Photographs
Photographs assist greatly in presenting technical evidence and help the justification of a technical case. They are also commonly included in technical reports because they bring some life to the text and can make quite 'dry' technical text easier to understand. The purpose of photographs is to *record technical detail* – particularly that which may be difficult to describe accurately without long and verbose wordage. Single photographs are of little use in a well-conducted engineering failure investigation – the necessary amount of information can normally only be captured by thinking of the investigation photographs as a complete 'set', each individual print holding a different, relevant piece of information to help build up the full picture. Figures 4.8 and 4.9 show a typical photograph set for an engineering failure. Note the following points:

- Figs 4.8(a) and (b) show typical 'overall view' photographs. You should take several of these, from different angles, to show the general arrangement. Note how the items fill the frame.
- Fig. 4.9(a) is a long close-up view of the failed component. Notice how the focus distance here has been chosen so that the failed component fills the frame – there is no room for distracting background items. This is approaching the limit of close-up resolution obtainable with a stan-dard camera. It comprises an oblique view of the fracture surfaces (ideally), generally known as the fracture *face*. The scale is particularly important in this photograph, as is the flash-lighting to obtain a clear picture. It is often beneficial to take several similar photographs of the fracture faces, each from a slightly different angle, to ensure the best view is obtained of the characteristics of the surface.
- Fig. 4.9(b) shows a magnified 'macro' view, in this case to show the existence of particular surface defects.
- *Annotations*. It is important to annotate all the photographs. This is to highlight points such as assembly details, direction of loads, regions of crack initiation, propagation and final fracture (we will discuss this further in Part II). Annotations draw attention to these points and prevent viewers looking too hard at the other, less relevant, parts of the photograph (which may convey different or incorrect conclusions). Remember that every photograph should have a scale in the picture, to show how big the components are.

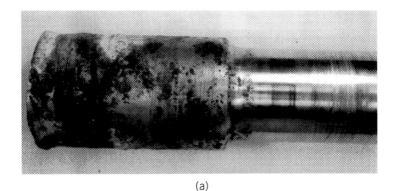

(a)

(b)

Figure 4.8   Overall view photographs (Courtesy of Royal and Sun Alliance Engineering)

The final type of photograph is more specialized. The *macrograph* and *micrograph* show enlarged views of the material and its fracture faces and are used to diagnose specific types and stages of metallurgical failure mechanism. They can sometimes be conclusive, and sometimes not – it depends on the type of failure. Both these photographs need separate camera microscope attachments, and/or specialized scanning equipment to obtain the images, and so are normally performed by specialist laboratories.

Action lists
Action lists are useful in several parts of the failure investigation process. They have a role as 'self-reminders' for the parties concerned

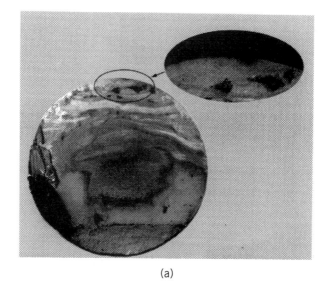

(a)

(b)

Figure 4.9   (a) Typical view of a fracture face (Courtesy of Royal and Sun Alliance Engineering). (b) A 'macro' view (Courtesy of Royal and Sun Alliance Engineering)

but their main use is normally as part of the outcome from a meeting or discussion. There are two types of 'action lists' – known generally as 'open' or 'closed'. *Open* lists contain open questions and only ask for similarly open-ended answers. The actions mentioned are often imprecise, and are included mainly to try to show that the parties are

(a) An 'open' action list

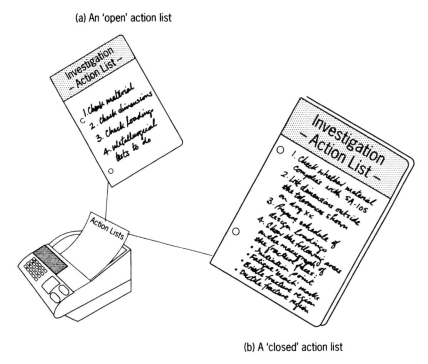

(b) A 'closed' action list

Figure 4.10   'Open' and 'closed' action lists

'doing *something*'. The result tends to be answers which are incomplete, inconclusive and spawn a lot of other actions. We have all seen action lists like this. The list grows after each meeting until it attains an unmanageable size (50–60 points) at which stage the parties either lose interest or make a series of rushed activities to partially address their actions – once again, their main objective is to show that they are doing something.

'Closed' action lists are much more precise instruments. The recorded 'questions' are phrased in a way that encourages tight, clipped, technical answers that *mean something*, and which can push the damage investigation forward, rather than leave it suspended in a technical limbo. Figure 4.10 shows examples of open and closed action lists – note how the questions in Fig. 4.10(b) encourage the types of constructive technical answers I have described. Please try to avoid using the types of action-list entries shown in Fig. 4.10(a). They may lead to the clear actions and answers that you want, eventually, but they also may not – and you will waste an awful lot of time and money in the process.

Interviewing

Interviewing is almost a technical subject in itself. In the course of an engineering failure investigation, interviews are an important source from which to gather information. Even in a simple failure case, there will be much more *relevant* information held by the various people involved (operators, designers and manufacturers) than exists in bland written form – formal documentation can often be dated or incomplete, and therefore misleading. The problem with obtaining information from people is that it is rarely presented to you in useable form. It may be interspersed with other interesting (but not directly relevant) bits of similar information, and it is almost certain to be a heady mixture of fact and opinion, some of it authoritative and informed, and some not. So, the difficulty with interviewing is:

IT IS DIFFICULT TO OBTAIN OBJECTIVITY.

As a rough guide, only about 20 percent of what you will be told during interviews can be treated as being truly objective. This in no way reflects on the integrity of your interviewees, rather it is a quality of the nature of information itself, and the way that people see, and remember, things that happen to them. This does not pretend to be a book about the psychology of interviewing: it would take a good few chapters to analyse the 'transactional' issues that take place between interviewer and interviewee. Instead, we will look at a few general *guidelines* that you can use to help your understanding – they are not absolute, or perfect, but I think that they have a certain kernel of truth about them.

Expect to find that, when conducting failure investigation interviews:

- Everyone believes the failure was *not their fault*.
- Nearly everyone finds it easy to believe, with embarrassment, that *another party* is at fault. You will be surprised at the gentlemanly ways some people find in which to accuse each other. Expect to find this convincing at first (until you have heard it a few times).
- In general, people *will* tell you what they feel to be the truth and will rarely set out to deceive.
- You will encounter embarrassingly short answers and a certain amount of well-timed diplomatic *silence* when your questions are probing uncomfortably near the truth. If you can recognize this, you are well on the way to finding the actual causes of failure, and promoting a settlement. It is easy to see, with a little practice.

Given these four observations you can perhaps anticipate the difficulty of obtaining true objective answers when interviewing people involved in failure investigation cases. One way to increase your chances of getting quality answers is to think of each interview as being divided into two parts, an initial broad *focusing* part and a final more precise *targeting* part. The purpose of the focusing part is to develop a feel for the scope of the technical issues inherent in the failure and so to develop a focus on the important ones. The technique here is to use 'open' questions, to allow the interviewees to give broad and developing answers. You are looking *only* for clues as to the real technical issues. Once you have this focus you can proceed to the second part where the idea is to concentrate on the key technical areas you have identified. You need questions here that are pointed, and which demand precise technical answers. You are not 'opening up' the issues, you want to tighten them, to close them down to encourage good quality answers. Remember that the whole purpose of this exercise is to home in on the ultimate targets of the failure investigation that we saw in Chapter 2: decisions on *causation*, and information that will promote a final *settlement*. This is quite different from trying to find a 'perfect answer' or promoting further expensive technical research work.

The timescale of interviews is also important. Your best chance of obtaining objective information lies at the beginning of a failure investigation, i.e. at your first meeting with each interviewee. People become more guarded, less objective, and remember less about the incident, as time progresses. Aim to arrange your interview early, try the two-stage approach I have recommended, and the results will start to come. You may find Fig. 4.11 useful in helping you structure your two-stage approach to interviewing. It shows typical examples of questions – you can soon differentiate them yourself, with a little practice.

These basic skills, combined with the type of tactical approach I have suggested, can help you carry out failure investigations in an effective way. Remember, though, that they are only *part* of a successful strategy towards investigating failures, they are not *substitutes* for it. The basic strategic approach, as shown in Fig. 4.1, is based on getting the correct focus on the investigation. This really is a vital point; if you take the wrong focus, then even the best tactical understanding and well-developed basic skills will not necessarily help you. The conclusions of a failure investigation are the important part – and you won't obtain the right conclusions if you lead in the wrong direction. Try hard to understand your, and others', focus on events, and you are well on your way to becoming competent at failure investigation.

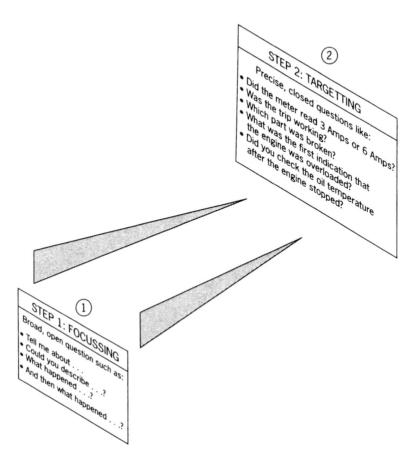

STEP 2: TARGETTING

Precise, closed questions like:
- Did the meter read 3 Amps or 6 Amps?
- Was the trip working?
- Which part was broken?
- What was the first indication that the engine was overloaded?
- Did you check the oil temperature after the engine stopped?

②

STEP 1: FOCUSSING

Broad, open question such as:
- Tell me about . . .
- Could you describe . . . ?
- What happened . . . ?
- And then what happened . . . ?

①

Figure 4.11   Interviews need a two-stage approach

KEY POINT SUMMARY: STRATEGY AND TACTICS

1. Strategy

You need to be properly organized to be effective during failure investigations. This means you need to understand the *strategy* of what you are doing.

2. Focus

Just about everyone in a failure case (engineers, users, insurers, designers, manufacturers) sees the nature and purpose of the investigation in a different way. Each party has a different *focus*.

3. Tactics

The basic tactics of good failure investigations can be summarized as:

- clear thinking
- logical progression (working in logical technical steps)
- talking up and down ⎫   (Techniques to help your case
- moving the argument ⎬   but without misrepresenting
- holding on ⎭   technical facts.)
- avoiding circular arguments (mainly about engineering design).

4. Basic skills

There are a few basic skills needed: collecting relevant information, making accurate notes, taking the correct photographs, compiling action lists, and interviewing people. These are all learnable.

Part I
Rewind

'That's rather an insult, my failure investigations don't end up in a mess'

'I suppose they're all different, but I've certainly seen a few that never seem to reach any conclusions'

'Good for you ... but why do I need this ... this model, the framework of damage and causation?'

'Maybe it's ...'

'I've got it, it holds the various concepts and ideas that make up the process of investigating failures – and then, well it's down to ... the engineering I suppose ...'

'Opposition'

'What?'

'Opposition; opposing views; offence and defence'

'Hmm, strong stuff, that's not the kind of engineering I'm used to, we all work together to solve problems – we're all professionals you see; consensus, solidarity, unity of technical purpose, that's what engineering is all about'

'And the interface?'

'The position of neutrality between opposing technical positions, in Figure 3.1?'

'He says it's empty – it has to be, by definition, irrefutably, empty – there's no-one there. No-one can be'

'I don't feel comfortable with that – like I said, there has to be an agreement on causation, before you can have a settlement; his words not mine'

'That's your comfort zone, you see – you'll have to learn to step outside it'

'I'll not be pushed – I'll go along with the professional responsibility and 'duty of care' thing, but I'm not arguing about money, that's for accountants'

'Have you tried, or is money outside your famous comfort zone as well?'

'Now you're saying it's all about money, it's . . .'

'Actually, it's only . . .'

'Argument, the icy blast of opposition, and the insistent love of money – what about the engineering?'

(Softly) 'Facilitator'

'What?'

'Facilitator . . . money is nothing more than a facilitator, it just helps you think about the way in which investigations are eventually settled'

'It's part of the engineering world?'

'That's up to you to decide, but it's part of the focus of an investigation, that's for sure'

'And that bit in Chapter 4, the 'clear thinking'?'

'Helps keep things simple'

'Who are you calling simple?'

'All of us, compared to the complexity that you can get in a failure investigation, if you don't keep it under control'

'Hey, you're having an effect on my comfort zone; money doesn't necessarily mean argument, and offence and defence don't have to infer either disagreement or offensiveness'

'Engineers one and all'

'Tough task though, making decisions about engineering failures, causation, settlement – and all that'

'Honestly, yes, if you want to do it properly'

'Any more assaults on our comfort zones in Part II do you think?'

Part II

Introduction

Part I of this book introduced the various concepts and principles that make up the subject of engineering failure investigation. Engineering failure cases are highly individual, but the fundamental differences are mainly technical; the 'building blocks' of the failure investigation process remain much the same. The *depth* of analysis in a failure investigation is an important consideration; in Part I we did not explore any of the cases in much technical detail. Part II will address this – it is intended to show how the 'building blocks' of Part I are used, together, in the context of a complex engineering failure.

Chapter 5 will look at the initial inspection that follows shortly after an engineering failure has occurred. The emphasis is on the practicalities (and sometimes difficulties) of the inspection, and preparing to fit your preliminary findings into the damage/causation framework introduced in Part I. In Chapter 6, we look backwards in time – I have tried to show a practical way to do a design evaluation on the failed component and then to look for clues in its operational history. These are early steps on the road of finding the causes of failure. They are both difficult stages, because they involve understanding and assessing the actions of other engineers, operating at both individual and company level. Chapter 7 continues the process of forming your technical opinions – it shows how to collect technical pointers and work towards identifying an initial *categorization* of the failure mechanism. Chapter 8 examines these initial conclusions and hones them down to a form in which they can be used to define *causation*. Defining causation can be a difficult exercise – the ideas introduced in Part I are used to structure the approach. Chapter 9 is a long chapter designed to show you (hopefully) how to present *short* conclusions. Good conclusions and clear descriptions of failure events and causation are the commercial 'product' of failure investigations. The chapter introduces, also, some of the protocol in this area, showing what you can and cannot present.

I made the statement at the beginning of the book that you don't necessarily need to read Part I in detail before working through Part II.

This still applies – you should find Part II consistent in itself although you may need to refer, in a very few cases, to Part I to clarify the formal meaning of some of the terminology. Part II contains a detailed case study showing the various principles in use. There is nothing contrived about this case study, it has similar characteristics to many failure cases of rotating mechanical equipment.

Chapter 5

Getting started – the inspection visit

Remember the statements in Part I of this book about having to step outside your engineering comfort zone? One of the best examples of this is your first inspection visit of a failure investigation.

When a piece of engineering plant suffers a serious failure the immediate steps are generally to isolate the plant and prevent any further actions that could confuse, hinder or otherwise interfere with the forthcoming investigation. Put simply, activities are put on hold while everyone tries to decide what to do. There is, therefore, a certain amount of suspense that builds up about the first inspection visit by 'the investigators', regardless of whether they are representatives of the insurers, the Health and Safety Executive, consulting engineers, or whoever.

The first inspection visit is when things start to happen. The ultimate success of your failure investigation, whether it converges cleanly towards a neat technical solution or diverges wildly, depends heavily on the seeds that are sown during these early stages of the investigation. Ideally, the inspection process would be leisurely: plan a little, inspect a little, plan a bit more, inspect again, re-visit to check that tricky conclusion, supplemented perhaps with a period of technical reflection for good measure. Sorry, but this all belongs to the comfort zone – in practice you won't have the time to do it like this, inspection visits are carried out within limited time frames. The other problem is one of *complexity*; engineering failures are almost always complex, containing several different engineering disciplines, and can involve large numbers of companies and individuals, each with their own interests to protect. Your eagerly awaited inspection visit is the start of what can be a long and (for some) rather disappointing path – so you shouldn't expect this visit to be *easy*.

Preliminaries – preparing for the visit

One of the secrets of effective failure investigation is *focus*; staying as close as you can to the core engineering issues of the case. You won't do this unless you have a reasonable feel for a set of objectives, before you start. This is an issue of style. It is about the way that you conduct an investigation – as if you mean to finish it, with a conclusion, an *answer*. So, you can set the earliest objective of a failure investigation before you even think about starting the inspection visit; here it is:

REMEMBER WHAT YOU ARE TRYING TO ACHIEVE.

Maybe now is the time to reconsider whether you really want to read this book – if the point above is so far outside your comfort zone that you can't bring it to bear on literally every activity of a failure investigation, then the subject is not for you, or you may be more comfortable with the other books shown in the bibliography.

The atmosphere to expect
This is something that needs consideration before you embark on the first inspection visit. Sadly, despite the suspense (or frustration) that probably precedes your arrival, you are unlikely to be treated as the all-conquering technical hero, freshly arrived to miraculously 'solve' the failure. Your welcome can be anything from openly cordial to overtly hostile – over time you will see, and feel, the full range. All engineering failures have an undercurrent somewhere of implied blame, so you can expect tensions. Few, if any, technical careers in plant design, manu-facture or operation have been built on the back of failure incidents. Thinking forward to the inspection visit, one of your important tasks is going to be to try to bypass any tensions that do exist (and you can't *yet* see them, remember) by bringing a strong 'fact finding' ethos to the inspection visit. This will make it *productive*. There are five guidelines to this; most of which you can't start to use until the visit itself, but it is wise to make them part of your preliminary thinking.

You can plan to improve the 'atmosphere' at the first inspection meeting by:

- Not taking *obvious* sides.
- Emphasizing the 'engineering' role in your activities.
- Keeping offence and defence (see Part I) until later.
- Forgetting the concept of victory and defeat (also in Part I), at least for the moment.
- Keeping at the back of your mind that objectivity is not the same as neutrality.

These guidelines are not perfect, or absolute, but they can help the situation. Build them into your 'comfort zone' and you will start the first inspection visit off on a sound footing.

Collecting background 'case' information

It is important to obtain information on the background to the failure case, before you attend the first inspection visit. Be careful not to confuse this with the collection of purely technical data relating to the engineering equipment or plant that has failed – the 'case information' that you need is almost purely *procedural* rather than technical. The most important information is that of the roles of the parties involved, either directly or indirectly. Essentially, you need to determine who is doing what. Fortunately, this falls into a common pattern, although the detail can differ between industries.

Figure 5.1 shows a general model that is applicable to many engineering industry sectors. The top half represents the parties involved in the *use* of the plant or equipment – the failure 'happening' will lie just above the chain-dotted line, normally within the jurisdiction of those operating the plant. Look how the other parties 'involved' above the line are shown as spreading into an inverted pyramid – this is simply indicative of their more general overall role, rather than having any significance about the level of responsibility that they assume. Note one key point about this hierarchy, though: it shows who is *nearest the failure*. This point has important implications for the way in which we conduct the failure investigation, because the party nearest the failure is virtually guaranteed to possess the best information about that failure. They will know what really happened, and will be able to contribute some of the harder pieces of the failure investigation 'jigsaw'. A further look at Fig. 5.1 will show that a similar, almost mirror-image situation exists below the chain-dotted line – these are the parties responsible for the design and manufacture of the failed plant. We must consider both halves, or we will only be getting part of the potential picture. It is not wise to attend the first inspection visit of a failure investigation until you know, broadly, which companies and individuals inhabit the various roles in Fig. 5.1. They will not, of course, all be present during your visit (although some will be). What you can be sure of, however, is that all these parties will, sooner or later, have an interest in the proceedings, and in how your findings will affect them. This means that you have to know, as part of your background information, who the people that you will meet are, and which interests they represent.

The other piece of background 'case' information that you will need is a broad idea of the *nature* of investigation that you will be conducting.

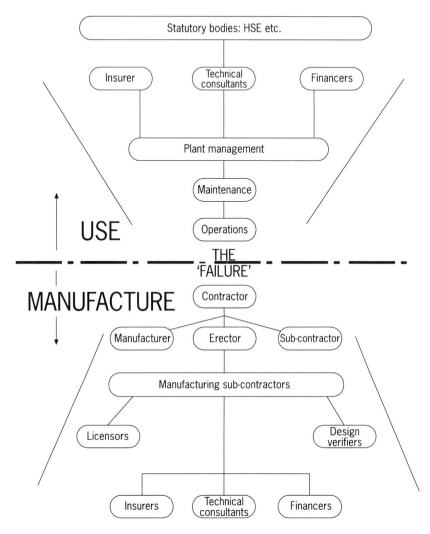

Figure 5.1   Background 'case information' – finding out who does what

There are four main types (A, B, C, and D) relating to pure commercial, liability, accident and research-orientated types of investigation, respectively – as introduced at the beginning of Chapter 2 of this book. Remember that each of these categories has a common technical core, but the investigation approach, and the emphasis placed on the various activities, differs a lot depending on which type of investigation it is. Try to get a reasonably clear view of this as part of your preparation for the first inspection visit – that way you won't either waste time or, worse still, have to repeat some of the investigation activities. In practice the nature of investigation is such that retracing your steps properly with a different 'type' emphasis is often impossible. If, after your first inspection visit, you suddenly appear again asking a totally different set of questions, your credibility will suffer, and it may never recover. So make sure you know which type of investigation you are expected to conduct before you visit. You can find this out by looking carefully at your brief, and asking for clarification, if necessary. You could ask for your brief to mention specifically one of the A to D types. This will help your focus, and encourage your principals (whoever they may be) to think. These are both good things.

Collecting background technical information

You will need this to complement the 'case' information. Although each case is unique, and hence the background and context of each plays an important part, the activities of failure investigation are still more than 90 percent *technical*. It is predominantly your engineering knowledge that will decide the outcome of the investigation – tempered only by the need to understand the correct approach – rather than your skills in negotiation, or debating, or administrative procedures. It is best to start collecting technical information as soon as you can, before you visit. This way you will understand a little about the technical aspects before you go. I cannot emphasize this too strongly; because the business of failure investigation is littered with the bones of companies – and people – who didn't know quite enough about what they were looking at (until, perhaps, it was too late).

So what advance technical information do you need? There are two basic categories: specific details of the failed equipment or component (from its *manufacturer*) and more general 'generic' information *about the way things fail*, available from several different sources. Figure 5.2 shows a sample checklist, using the example of a failed bridge member. Note the relative amounts of information that I have shown: there is much less of the structure-specific manufacturers' information, perhaps because it might prove more difficult to get. Either way, make sure you allocate some time to both of these information sources – you will need to use them together in the later stages of the investigation. One gentle reminder: you have to *go and look* for this information; sadly, it won't come looking for you.

Logistics – organizing the visit

This is really just about using common sense. You can't just turn up on the day, in someone's plant or construction site and expect that

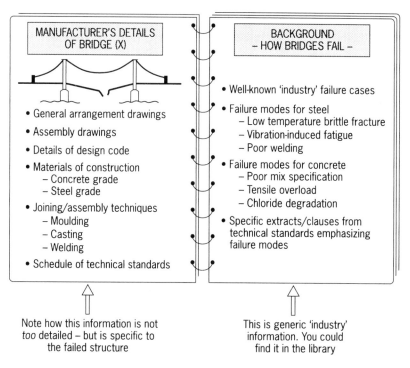

Figure 5.2   Advance 'technical information' you need – an example

everything and everybody will be prepared and waiting, just for you. Even the simple act of inspecting visually a piece of failed plant will involve arrangements for access, safety considerations and the almost inevitable 'management approval'. All of these take time and sometimes an unbelievable string of management meetings, only to eventually agree that you can actually come and look at the failure, so:

- Arrange your visit with a little bit of formality. Make sure people know *who* you are, *why* you are visiting and roughly *what* you intend to do.
- Submit an agenda. Stick to procedural matters; say who you will meet and which items of failed plant you want to inspect – but don't go into *technical* details.
- Mention that you will want to look at plant records and relevant documentation.
- Observe the normal business courtesies of confirming your visit beforehand, agreeing arrival and meeting times, then sticking to them.

First contact – asking and listening

Your main role during the first inspection visit is asking and listening rather than talking and concluding. There are important differences. These early stages of an investigation, before the serious technical discussions start, are the best time to get to the real roots of what happened, before everyone starts to become defensive. The way to do this is by asking and listening, in whichever order fits most easily into the situation that you find. We looked in Part I at some basic tactical aspects of interviewing (take a quick look back at Fig. 4.11, showing the two-stage approach to asking questions). Use all these techniques now, during the early periods of your failure inspection visits.

Start slowly

A golden rule is to start slowly. All the explanations of what the failure is, and how it happened, will not come at once. They always, *always*, take time to develop, more often than not outside the time frame of the initial inspection visit. You need to be prepared to let the story develop. Develop it will, so:

Don't jump to instant conclusions at the first inspection visit – give the *better conclusions* time to develop.

Another rule to follow during the inspection visit is to spread your asking and listening net as widely as possible. Plan to listen to all levels of operators', engineers' and management's versions of the failure story, as they see it. They will not, of course, all be totally accurate; management views, particularly, tend to be rather 'rarefied', usually underpinned by the eloquent assertion that, whatever it was that happened, it clearly was not their fault. Similarly, operations staff can often have a view of events slanted towards previous occurrences that happened somewhere in the past. Expect to hear tales of bravado, and reports of great feats of analysis, and others' indecision. You have little choice but to accept all this as part of the complex nature of engineering failure investigation (which it is). Don't forget the one clear fact, however, which stands above the milieu – you have seen this before:

Those people that were *nearest* the failure, in an organizational as well as a physical sense, will understand best what really happened. Ask them.

If you follow this general approach it will help you make useful contacts, on a technical level, with the people (from the parties shown in Fig. 5.1) that you meet during the first inspection visit.

The discussion route

The route that discussions will take is an absolute function of whether you are dealing with a type A, B, C, or D failure investigation. The broad principle, however, is general and is perhaps best explained by reference to Fig. 3.1, showing the general model of a failure investigation as a process of interaction between the investigat*or* and the investigat*ed*. Whatever the type of investigation (except for maybe the purest form of 'research') the discussion routes will be contained within this confrontational model. In practice, most of the early discussions will be about straight technical subjects, such as design, manufacture and operation, rather than about procedural matters like responsibility and blame. They will come later. As technical investigator, it makes sense to *fit in* with the early discussions – you can only do this by becoming actively involved, so that you encourage a free interchange of technical ideas. This is one area that causes difficulty for some failure investigators, even experienced ones. It is all too easy to think that becoming involved in technical discussion can weaken the investigator's position, because it can reveal technical insecurities. I am not sure that this is true – but what *is* certain is that if you don't discuss a failure, and actively expose its technical facts, your chances of ever understanding the real cause of failure are pretty small. Draw your own conclusions.

Expect some randomness, even confusion, in these early stages. The technical meetings that occur during the first inspection visit will *not* be logical highly structured affairs, with a neat set of conclusions and finely crafted points of agreement. Even a simple failure with an apparently obvious 'cause' can, and will, be interpreted in several different ways – and have hidden technical facets which will not appear immediately. This is such a definitive property of failure investigations that you need to be very wary of failures that *appear* simple and straightforward to solve. So:

Don't expect to 'solve' a failure investigation during your initial visit.
and
If it does look easy, you're probably not looking (or thinking) hard enough.

To avoid pitfalls, it is best to treat the early stages of the visit as a search for *reference points*. These are basic technical (and sometimes procedural) facts which are clear, and absolutely not in dispute. You can think of them as 'points of fact', although it is important not to refer to them as such, for procedural reasons that we will see later. Most reference points actually involve the process of *elimination* – a sound and useful engineering technique which helps the investigation onto a path of convergence. Figure 5.3 shows some examples of the types of 'reference points' that it is possible to identify during a first inspection visit – it refers to the case of a failed crane winch.

In summary, the first inspection visit, while it rarely produces the results of the investigation, is an essential preliminary activity. It sets the structure, tone and pattern of the later activities. Don't be worried if it *feels* a little random, you will be dealing with new events, and people, and probably wide-ranging technical subjects, not all of which will be totally familiar to you. This early character of the investigation will

---

Example: Failure of a crane winch

REFERENCE POINTS

- 'The winch drum works in an ambient temperature of 30°C' (this *eliminates* low temperature brittle fracture possibilities).

- 'The crane rope is 20 mm (minimum) diameter stranded steel' (*fact* – you can measure it).

- 'The steel rope is not broken' (*fact*).

Do not confuse proper 'reference points' with areas that are not so straightforward. For example:

- 'The winch-drum retaining bolts are undamaged' (this is an *assumption*, not a fact, until a full investigation has confirmed it).

- 'The crane design appraisal certificate indicates that the winch was correctly designed' (not necessarily a fact, and actually rather meaningless – what does *correctly* mean?)

- 'The crane was designed to BS 466/ISO 4301-1' (basically true, but an *incomplete statement* of fact – there are likely other design codes involved as well).

Figure 5.3 Some good (and not-so-good) 'reference points'

soon fade, and be replaced by a more structured and organized course of events as the technical issues unfold. This demonstrates a good point about engineering failure investigation:

> You have to wait for failure investigations to *unfold*. Don't expect instant answers.

Some early ideas

At some point during the initial inspection visit, you have to start to collect your early technical thoughts on what 'category of failure' you are looking at. At this stage, this is little more than a scoping exercise; a broad and probably rather crude assessment of the failure. The objective is to cut down the field of vision (for yourself, and others) to a manageable level. This will help everyone concentrate better on the relevant technical matters and help the *focus* of the investigation.

Figure 5.4 shows some examples of 'scopes' of failures. Note how

---

YOUR EARLY 'IDEAS' ON THE FAILURE COULD BE:

Mechanical rupture (something is *broken*)

or

There is excessive *wear*

or

Some parts are *missing*

or

This piece of equipment just *won't work*

or

The damage is due to *fire* (the pieces are burnt)

or

There is a *potential* future failure (due to cracking or yielding etc.)

Note how these are simple, 'lowest common denominator' ideas.

Figure 5.4   Initial 'scoping' of the failure – a simple start

general they are – there is certainly no attempt at this stage to define the cause of failure, or even the exact nature and extent of it, with any accuracy. It is more about deciding the visible consequences of the failure rather than the failure itself. You may feel that the broad categories shown in Fig. 5.4 are so obvious as to be of little use. In real investigations, however, you may find yourself surprised at the frequency with which parties cannot agree which of these categories they are discussing. I have seen many investigations where, after a few meetings, what started out as a failure due to excessive wear of a component turned into one involving a missing design feature, or a broken piece, or something which represented only a potential future failure. The end result of all this is an *awful* mess (and large amounts of wasted time and money). Can you see how an early exercise in categorization, using simple definitions, without their rigid technical interpretation, can help stop this happening? I am not saying that this initial step is easy, or that consensus will always follow, but if even only partially successful, it is going to make the investigation so much easier. There are a few simple guidelines that apply to this activity.

- It is only a scoping exercise, so don't use formal definitions – keep your thoughts and terminology loose.
- There is nothing wrong with expressing your thoughts to the other parties at this stage – but do it informally.
- Listen to what others think – you do not have a monopoly on technical knowledge.
- A word of caution: *never mention causation* at this early stage. If you do I can guarantee that your statements will be misunderstood, misconstrued, or both. Causation comes later.

From these points you can see that this stage is predominantly verbal – informal discussion between the plant operators, manufacturers and technical specialists, managers, technicians and interested observers. You can involve them all.

Next step – the investigation plan

Towards the end of your initial visit comes the time to start to think about your *investigation plan*. This will outline the approach to the main body of the investigation. It is a plan which is relevant to your own investigation activities, rather than something that you would disclose to the other parties. Investigation plans follow a common pattern in

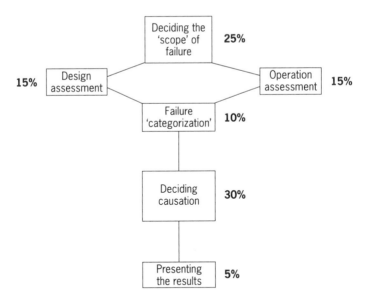

Typical (time) weightings shown for a
commercial/insurance-related failure investigation

Figure 5.5   A typical investigation plan

most engineering investigations, it is mainly the *emphasis* placed on each
individual activity that will differ.

Figure 5.5 shows the general format of an investigation plan. Note
the six main boxes: scope of failure, design appraisal, operation appraisal, failure categorization, causation and presenting conclusions. For
each investigation the relative amounts of effort put into each of these
stages is going to have a large effect on the outcome, i.e. whether it is an
effective investigation which reaches firm (and correct) conclusions, or
whether it becomes a time-waster. Your main task in specifying the
investigation plan (if you use the format suggested in Fig. 5.5) is to
estimate the percentages of *time* that you will allocate to each step.
There is no 'right' answer to this because every case is different. There
are, however, some broad guidelines that you can follow.

● Start by thinking carefully about how much time you will need to
  decide the scope of failure (the first box). If you don't allow enough
  time for this, you may only reach a snatched conclusion, and the
  investigation will soon develop into a circular, indecisive path of 'ifs
  and buts'. You need at least 15 percent of your budgeted investigation time to come to proper decisions in this area.

- The design appraisal and operation appraisal are rarely given enough planned attention, although they are frequently mentioned and discussed loosely. Effective investigation involves a *structured* approach to assessment of the design and operation of a failed component – so you can find out what influences they had on the failure. This is of particular importance in insurance and commercial liability-based investigations.
- Be careful about the categorization and causation stages. It is easy to spend nearly all your time on these, without any real results, if you haven't grasped fully the details of the scope of the failure. Remember the downside – the unproductive circular arguments (see Chapter 4) that await you if you get it wrong.

In general, the format of the investigation plan is more or less independent of the type of failure investigation (A, B, C, or D – take a quick look at Chapter 2 if you need reminding of these definitions) under consideration. There is, however, a tangible effect on the relative weightings of each of the activities in the plan. There are no hard-and-fast rules on this, only guidelines:

- Failure investigations with a strong commercial bias (type A) tend to concentrate more on the *causation* aspect. Insurance investigations, specifically, are almost totally orientated towards deciding what caused the failure, and whether this was an 'insured peril' (we will look at this in more detail later).
- Liability-based investigations (type B) are often more concerned with *who* was at fault, with less emphasis placed on describing the precise mechanism of failure or its cause. For this reason, the activities of design appraisal and operation appraisal often take priority – up to 70 percent of the investigation time in some cases.
- Accident investigations (type C) are perhaps the most 'balanced' category of investigation that you will meet. All aspects of the failure have to be seen to be investigated in equal and adequate depth, so you can expect a roughly equal apportionment of time and effort between the boxes in Fig. 5.5. All of the accident investigations that I have seen have followed this sort of pattern, but they haven't all produced good, clear answers. Follow your own instincts on this one.
- Type D investigations, as explained in Part I, are research- or technology-based inquiries in their purest form. They rarely have dramatic conclusions. Frankly, it can be difficult to tell exactly which activity they are concentrating on, because they sometimes lack focus and are steeped in technical and procedural rigour. The more

focused ones tend to look closely at the *cause* of failures, rather than the other aspects.

So what exactly do you *do* with your investigation plan? Its main function is to help your focus – to help remind you about the format of the failure investigation, so that it will be effective, and converge, and reach conclusions. These are fine principles. On a more practical level it is a useful point from which to start making *checklists*. Checklists are important working tools in failure investigations – they are a good way of ensuring that you marshal your engineering facts and technical thoughts in a structured way.

You can think of the investigation plan and its checklists as one of the main *outputs* of the initial inspection visit. We can crystallize this to give a clear brief for the visit:

• Your input to the visit is looking, asking and listening.

and

• The output is a clear investigation plan, and an accompanying set of checklists.

Checklists

What should you put in your end-of-visit checklists? The main content is simply a list of things that you need to find out, nothing more complicated than that. This is also not the place to start to write notes about things that you know already. It should stand alone as an action list to give you guidance once you get back to the office. The best way is to divide your checklist into six parts, mirroring the six boxes shown in Fig. 5.5. Figure 5.6 shows a sample (although simplified) checklist taken from a type A investigation into a crane rope failure. Note the format and the relative emphasis on each of the six areas.

Summary – before you leave

The first inspection visit normally proves to be the best time to get easy access to clear and unbiased information from people involved directly in the failure. You can expect things to become harder as the investigation progresses, for reasons already mentioned. This means that you have to make a real effort during the initial visit to obtain the information that you need to *support* your investigation. Take this seriously – it may be your first *and last* visit to the actual failure site so you have to get the information you need *when you are there*. This means collecting data, photographs, drawings, system diagrams,

---

CHECKLIST (BROKEN CRANE ROPE)

1. Nature of failure; i.e. what actually happened?

- Check the list of damaged components – is it complete?
- Verify initial diameter of the broken rope (with manufacturer).
- Check the normal 'replaceable components' – what are they?

2. Design appraisal

- Confirm the crane design standard is ISO 4301-1.
- Obtain the crane's inspection certificate.
- Check the rope factor of safety (from standard or data sheets).

3. Operation appraisal

- Obtain Operations Department's statements about the failure.
- Ask about maintenance records for the brakes and crane rope.

4. Categorization

- Check the crane technical literature (to see if *wear* is common-place).
- Look at the standard on crane ropes (how are *overloads* prevented?).
- Check material standard (which grades of steel are used for stranded rope).

5. Causation

- Rope dimensions to record – to check for *overload* conditions.
- Rope manufacturer's QA/QC certificate to review – check the probability of rope *defect*.
- Do steel ropes deteriorate? – look at literature.
- Can rope strands fail *catastrophically* if overloaded? – check reference information.
- Could a jammed brake be a contributory factor to rope overload? – check with manufacturer.

6. Presentation

- General arrangement drawing of the brake assembly, needed.
- Macrograph photograph of a fractured rope (check availability).

Figure 5.6   A (simplified) end-of-visit checklist

material samples, operational records, names of equipment manufacturers; whatever you need to allow you to build a coherent picture.
Research work and detailed analyses can come later, back in the office.
It is the *technical information* from site that is going to be the key to
your investigation, so don't waste too much time in discussing
procedural or commercial matters, interesting though they may be.

Following the initial investigation visit the next step has a more
rigorous engineering basis. We need to look at the *design* of the failed
equipment and form some conclusions about the way that it has been
operated.

KEY POINT SUMMARY: THE INSPECTION VISIT

The *inspection visit* is the first step of a failure investigation. Expect to have only limited time. Keep an open mind about the kind of welcome you (as technical investigator) will get.

1. Background information

Collect background information *before* you go. You need:

- Background on the type of investigation you are required to do.
- Detailed technical information (about the failed equipment).
- Good technical background knowledge about failures.

You will have to *go and look* for this information. It won't come looking for you.

2. Asking and listening

The first inspection visit is mainly about asking and listening. Start slowly but keep going.

3. An early solution?

Probably not. It is rare to 'solve' a failure case during the first visit. You have to give things time to develop. Try to find the general 'scope' of the failure. Do a rough investigation plan (Fig. 5.5) and use checklists.

Chapter 6

Design and operation assessment

What do the assessment of the engineering design and equipment operation have to do with investigating failures? Shouldn't we be moving on quickly to talk about the distinct mechanisms of failure – looking in detail at the metallurgy? This comes later. If you think forward towards the conclusion of a failure investigation, where it is necessary to make firm statements about causation, then the issues of the *design* of the failed component, and the way in which it was *operated* before the failure, are of prime significance. You will find few failures where you will not be told, by someone, that the failure was due to 'the wrong design' or that 'it wouldn't have failed if it had been operated correctly'. There are indeed implications if such conclusions do prove to be fact – commercial and insurance liabilities are affected in various ways if inherent design faults or operator negligence can be proven.

The design and operation assessments are two of the rather more ordered parts of a failure investigation and are done shortly after the early inspection visit. They are predominantly based on technical facts and analyses rather than anything procedural or commercial. They are by no means easy analyses to do, but they do respond well to the structured and knowledge-based approach that many engineers bring with them. Perhaps the main barrier is that both the design and operation assessments involve looking *backwards in time*. The mechanical design of even the simplest component has evolved as a result of chains of design decisions made well in the past, many of which will not have been recorded. Equally, records about how equipment was operated are rarely perfect, particularly on older plants which are, statistically at least, the ones most likely to suffer mechanical failures. The sheer *complexity* of the design and operation assessments is a problem.

The greatest emphasis is on the assessment of *design* – engineering design is a multidisciplinary subject, full of specializations, experience, heuristics ('rules of thumb') and partly hidden standards and conventions. This means that, on assessing a design, you will be moving outside your technical specialization (if you have one) and towards the boundaries of your experience and general 'comfort zone'. Assessing plant operation is a little easier, but you can still experience difficulties.

Fortunately, most of the potential problems can be diffused by approaching the task of design and operation assessment in a structured way, and by taking careful steps to organize your activities. This chapter will look in some depth at how to do it, working off three general guidelines:

- Treat design and operation assessments as *separate* technical activities.
- Expect technical *complexity* (because it will always be there).
- Remember that you have personal technical *limitations* – so you need to keep your activities under control.

The easiest way to understand the principles of design and operation assessment is by the use of an example, so you can see the technique in action. The case study example used is that of a radial fan which has suffered failure of the main drive shaft. This is not the most straightforward example to use (a simple welded joint would be much simpler) but it is useful in that it is complicated enough to allow the development of a general framework of design and operation assessment without the analysis becoming too submerged in advanced technical detail. We will follow this example through the remaining chapters of the book to show the development of a typical failure investigation – its difficulties and (hopefully) a set of useful conclusions.

The design and operation assessments are both exercises in *elimination*. Their main purpose is to enable you to eliminate clearly those aspects of the equipment's design and previous operation that did *not* have any influence on the failure that occurred. In the course of the assessments you may find some design or operational aspects that *were* relevant to the failure. It is best to treat design and operation as the separate issues that they are – they need different trains of thought and benefit from different types of methodology:

- *Design assessment* is based on sound engineering knowledge, tempered by experience and a little intuition. A wide-ranging analysis is necessary, to look for design features that are relevant to the failure mechanism.
- *Operation assessment*, in contrast, is much less definitive. It has 'softer' boundaries and often involves some uncertainty about

what did or did not happen during the pre-failure life of a piece of equipment.

Both assessments are equally important parts of the overall failure investigation. It is worth mentioning again that perhaps 70 percent of their purpose is that of *elimination*; only for the remaining 30 percent will you be dealing with design or operation issues that are directly related to the failure. We need to look at them in some depth. First, the design assessment.

Design assessment

It is difficult to make *accurate* assessments of mechanical design because:

- your knowledge is limited (you are not the designer)
- the nature of the design process is against you – it is multi-disciplinary, iterative and complex
- time is limited.

You have to be realistic – you can only do an effective design assessment if you can simplify things down to a workable level of complexity. A simple framework that you can use is shown in Fig. 6.1. Here the subject of mechanical design is broken down into six basic elements (shown as boxes). These are the parts of the design activity that contribute most to the fitness-for-purpose of mechanical equipment – many of the less important activities (I could have shown them as smaller boxes) have been 'stripped out', leaving only the essential framework.

The methodology – how to use it
To use Fig. 6.1 it is necessary to develop it a little further to show some of the technical design parameters that lie within the six main framework boxes. I have shown this development in Fig. 6.2. Note how the technical parameters have been divided into two halves. Those to the left of the framework subjects are relatively *easy* to assess – they can be evaluated by using published design data and 'rules of thumb'. Those on the right side are more difficult, mainly because of the nature of the engineering disciplines – these can be complex (and confusing) and rely heavily on empirical design rules that have been developed, over long timescales, from experience. So, with this as background, how do you actually *do* a design assessment? Here is the first rule:

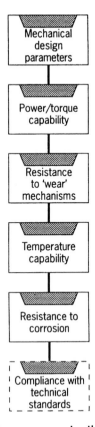

Figure 6.1   Design assessment – the basic framework

ALWAYS *START* YOUR DESIGN ASSESSMENT FROM THE *FAILURE.*

This gives you your starting point. By starting from the failure – both in physical 'location' terms and in relation to the various technical parameters shown at the sides of Fig. 6.2, you are starting to develop the analysis in a structured way. This is good *focus* (remember Chapter 4?).

Your objectives
You are looking for evidence of three things:

- *Design limits* – once you have identified the design limits of a component, you can find out whether the failed component was operating *outside* any of these limits.
- *Design inconsistencies* – mismatches of forces, component sizes, ratings, strengths, materials, hardness values and flexibility are all

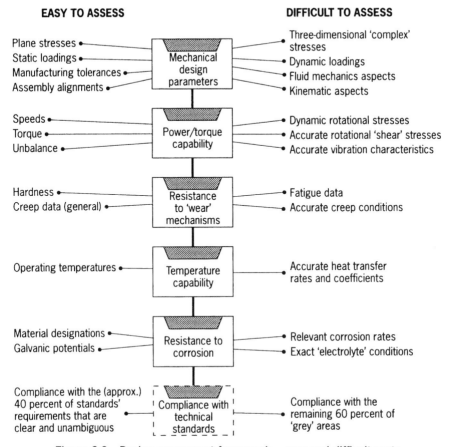

EASY TO ASSESS 　　　　　　　　**DIFFICULT TO ASSESS**

Plane stresses ●———————　　　　　●Three-dimensional 'complex' stresses
Static loadings ●———————   Mechanical   　●Dynamic loadings
Manufacturing tolerances ●———   design   　●Fluid mechanics aspects
Assembly alignments ●———— parameters   ●Kinematic aspects

Speeds ●———————　　　　　●Dynamic rotational stresses
Torque ●——————— Power/torque   ●Accurate rotational 'shear' stresses
Unbalance ●——————— capability   ●Accurate vibration characteristics

Hardness ●———————   Resistance   ●Fatigue data
Creep data (general) ●———— to 'wear'   ●Accurate creep conditions
　　　　　　　　mechanisms

Operating temperatures ●——— Temperature   ●Accurate heat transfer rates and coefficients
　　　　　　　　capability

Material designations ●——— Resistance to   ●Relevant corrosion rates
Galvanic potentials ●———— corrosion   ●Exact 'electrolyte' conditions

Compliance with the (approx.)
40 percent of standards'   Compliance with   Compliance with the
requirements that are ●—— technical   ●remaining 60 percent of
clear and unambiguous   standards   'grey' areas

Figure 6.2   Design assessment framework – easy and difficult parts

common examples of design inconsistency that goes against good design practice. There are many others. Mismatch can also happen on a large scale when, for example, machines such as prime movers or gearboxes are 'packaged' with inadequate thrust bearings, or too-rigid casings.

● *Non-compliance* with published design standards.

These three areas form the main 'hunting ground' of the design assessment. The technique is to work through the relevant technical aspects of the design, paying particular attention to these three objectives (in reality you can treat them as the *only* objectives). Remember that the objective is to *look for problems* in these three areas. It is easy to take the wrong view and to waste time looking for evidence that the failed components *did* have neatly matched design

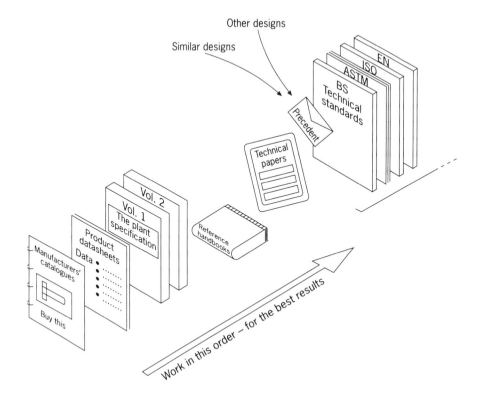

Figure 6.3   Assessing a design – how to use the information sources

parameters and careful compliance with the letter of design standards. Such conclusions are of no use to you – it is the poor margins, mismatches and non-compliances that you want.

Information sources
For the mechanics of the design assessment process, you need a good general engineering knowledge, coupled with the ability to obtain information that you don't have. This is an exercise in *selectivity* – engineering information comes in many forms and it is easy to become submerged in technical details. The skill is in selecting only the information that you need, no more. Fortunately, there is a well-defined hierarchy of source information that you can use as a general guideline. This is shown in Fig. 6.3. Don't regard this as being anything to do with the activity of actually designing a piece of equipment – it is related strictly to the field of *assessing* an existing design. This is an important distinction. Look how the figure shows a hierarchy of seven sources of

technical design information. Using the information sources in the order shown is the most effective way of working through the mechanics of the design assessment with the least chance of wasting time or effort. Surprisingly, this order of approach works well for many different mechanical equipment items, and, on a larger scale, with process and system designs. It does not work well with electrical or electronic equipment, because their design process is very different. The approach can be seen in use later in the fan failure case study. We can now take a closer look, in order, at each of the information sources:

- *Manufacturers' catalogue data.* This includes the data sheets for the failed equipment or component. Data sheets are produced for most types of manufactured equipment and are the best place to start the design assessment. Data sheet information is expressed in a brief but comprehensive way, which makes it easy to identify design limits and technical standards – two important parts of the design assessment objectives. Data sheets are commonly included in the equipment's instruction or operating manual – if not, then contact the manufacturer, quoting the component or equipment type and serial number. I cannot emphasize too strongly this need to start a design assessment from the relevant data sheet – it is vital technical background.
- *Manufacturers' product support literature.* This is a good second step. Equipment manufacturers publish good generic technical information articles and 'papers' about their type of equipment. This can contain useful additional information on design limits, matching and relevant technical standards (our three objectives again) in a way that is useful to a design assessment. Information available from manufacturers of machine 'elements' (bearings, gears, shafts, pins, couplings, etc.) is particularly well detailed – and these are often the items that fail. Expect to have to ask manufacturers for this information – it probably will not be in the instruction manual or general product catalogue.
- *The plant specification.* This means the technical specification for the equipment that has failed. While it may not be particularly detailed, it will contain relevant information about design parameters (temperatures, pressures, speeds, loading regime, etc.) that influenced the size and type of equipment chosen, so it is relevant to your assessment of design limits. This specification also has important contractual relevance to whether the equipment was fit for its *specified* purpose.

- *General engineering references.* Don't be misled by this terminology. 'General' references provide technical information that is not necessarily product-specific – which does not mean that it is not directly relevant to the failed equipment. The best example of this is data contained in Engineering Handbooks such as Kempe's, reference (**1**), and Marks', reference (**2**). Both contain good quality information which can be used during a design assessment. In the right context, they provide useful technical back-up to manufacturers' product-specific data.

- *Authoritative technical sources.* This is rather a 'catch-all' category, comprising technical articles, papers and material from other published sources. Your main problem will be deciding exactly how relevant such information is to the equipment under assessment. Various well-known 'generic' failures are often analysed in some detail in specialist technical articles, but you will be lucky if your assessment falls squarely into one of these areas. Expect to have to pick and choose carefully in order to extract really useful design information from specialized technical publications.

- *Design precedent.* This is an interesting one. Theoretically, the findings of design precedent should be absolute: i.e. if a component (x) has been proved to work successfully in a particular piece of equipment (y), then this proven combination of (x) in (y) should act as a precedent of fitness-for-purpose, whenever x and y appear together in a design. In practice, it is rarely this simple – the variety of detailed design parameters and operating conditions for any piece of equipment is so wide as to negate the integrity of the precedent. The fact that a particular design of, for example, a shaft bearing is proven in one type of pump does not mean it will necessarily be suitable for use in other, even similar, types of pumps. Look for design precedent, by all means, but be wary of using it as a core activity of design assessment. It is often weak, and sometimes seriously flawed.

- *Technical standards.* Published technical standards are the most *complex* source of design assessment information that you will encounter (that is why they are shown at the end of the hierarchy of sources in Fig. 6.3). Paradoxically, they also hold the most detailed and targeted information, and present it with a certain degree of authority. There are two fundamental types of technical standards; those primarily related to *design*, and those mainly to do with the activities of *manufacturing*. There are also those which are hybrids – a mixture of the two. All three have their place but it takes a little experience to know exactly which is which. Don't make the awful mistake of relying on, or trying to quote from, a technical standard which you have never read. You can see this happening quite often in design assessments, particularly if there is time-pressure to get on with the failure investigation. Imperfections apart, technical standards are good sources of data on design limits and tolerances – information which you *will* need to identify any design mismatches or inconsistencies that may be relevant to a failure.

The above sources cover just about all the 'external' technical information you will need to compile the design assessment. This information does not stand alone – it *adds to* the direct information obtained from your initial inspection of the failure. Design assessment is predominantly a desk-based exercise but it has to be done in context with the findings of the inspection visit, not in isolation.

Design assessment – an example
The best way to understand the principles of design assessment (and most of the other elements of engineering failure investigation) is by using an example. Figure 6.4 introduces the failure case study that we will follow through the investigation stages, culminating in the conclusions on causation and commercial settlement at the end of the process. Figures 6.4(a) and 6.4(b) form the first introduction to the failure case – they contain the typical level and depth of information that may be presented to you, most likely in verbal form, as the first notification of your technical involvement in the failure investigation. Look at the generally superficial nature of the information, and the way in which it is presented – it has different levels, inherent assumptions, and quite a few gaps. Figure 6.5 is a reminder of the recommended *approach* to the investigation. Before proceeding with the design assessment, we need the results of the initial inspection visit. Figure 6.6 is a summary of the inspection findings made during this visit, exhibited on the manufacturer's drawing of the fan, taken from the on-site maintenance manual. You have taken care to ensure that the drawing is of the exact fan model that failed. You have chosen the three-dimensional drawing because it is more descriptive, and easier to work with, than formal orthographic drawings. Figure 6.6 contains all your on-site observations about the fan's failed components. It is absolutely clear that the shaft is broken but you have also found wear, corrosion, and some loose bolts and have received a report (during a discussion with one of the operators), that this fan had a history of 'excessive vibration' and 'often seemed to be

---

DETAILS OF FAILURE

- A large radial fan, supplying air to a chemical process, has failed.
- Some valves were found to be shut.
- The failure resulted in a total plant shut-down, causing solidification of 1500 tonnes of process liquor and consequential loss of output.
- The user claims there is an insurance policy in force.

THE EQUIPMENT

- Electrically driven 250 kW radial fan approx 19 months old. There are five more in other parts of the plant.
- Manufacturer: The Radial Fan Co Ltd.
- Type: RF-1 (see users' sketch: Fig. 6.4(b))
- Location: mounted on a plinth outdoors, approx 10 metres above ground.
- Technical standards: BS 848 and relevant vibration standards.

INVESTIGATION BRIEF

- What caused the failure?
- Whose fault was it?
- Are the other five similar fans safe to use? (The user is concerned that this might happen again).

Figure 6.4(a)   The fan failure case study – a first introduction

overloaded'. Another operator thought the fan was 'under-designed' but didn't (couldn't) explain exactly why, or how. Now for the design assessment:

Step 1: Check the mechanical design parameters
Figure 6.7 shows a rough analysis of the mechanical design parameters (introduced in Fig. 6.2) applied to the radial fan. This type of analysis is usually done intuitively, identifying those mechanical design parameters most relevant to the observed failure(s). Preliminary observation of the failure suggests that two areas are worth looking at:

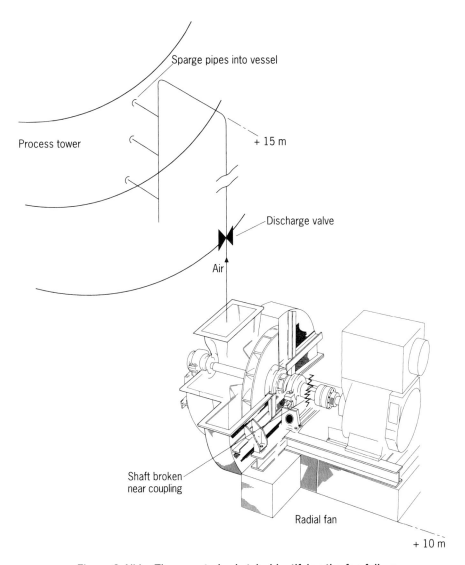

Sparge pipes into vessel

Process tower

+ 15 m

Discharge valve

Air

Shaft broken
near coupling

Radial fan

+ 10 m

Figure 6.4(b)   The operator's sketch, identifying the fan failure

- the shaft key loadings (because the shaft break is near the end of the
  keyway)
- manufacturing tolerances for the shaft-to-coupling fit (to see if this
  could be relevant to the shaft breakage).

Can you see the method behind the choice of these two areas? They both
have a direct relationship to the shaft break and are near enough (in

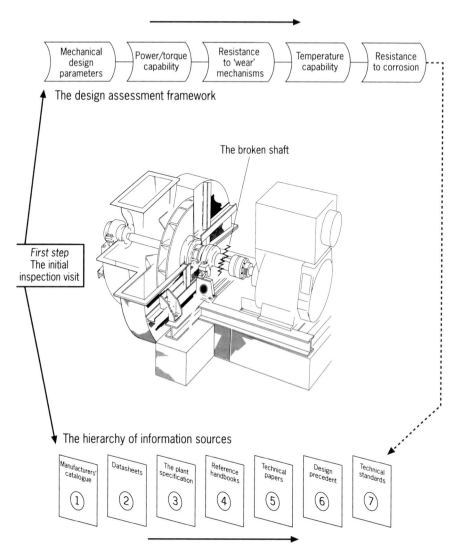

Figure 6.5   A reminder – the approach to your design assessment

design terms) to the failure to suggest that an assessment of their design could help provide useful *pointers* towards the nature and cause of the break. Homing in on the correct areas of design assessment is one of the key skills of successful failure investigation. You have to apply engineering knowledge, judgement, *and* experience. If you get it wrong, you can waste large amounts of time assessing irrelevant parts of the design.

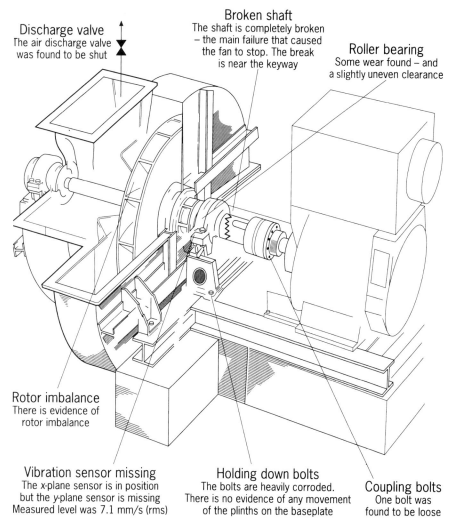

Discharge valve
The air discharge valve
was found to be shut

Broken shaft
The shaft is completely broken
– the main failure that caused
the fan to stop. The break
is near the keyway

Roller bearing
Some wear found – and
a slightly uneven clearance

Rotor imbalance
There is evidence of
rotor imbalance

Vibration sensor missing
The x-plane sensor is in position
but the y-plane sensor is missing
Measured level was 7.1 mm/s (rms)

Holding down bolts
The bolts are heavily corroded.
There is no evidence of any movement
of the plinths on the baseplate

Coupling bolts
One bolt was
found to be loose

Figure 6.6   The fan failure – this is what you found

Figures 6.8 and 6.9 show the design assessment activities for the key size and the limits and fits – in this case comprising straightforward calculations. Note the following 'principle' points:

- They are *'order of magnitude' only* – there is little point in calculating down to fine levels of accuracy.
- They use simple, *well-accepted formulae* for plane and torsional shear stresses. Trying to anticipate complex three-dimensional stress fields belongs to engineering research, not failure investigation.

Design parameter	Relevant fan component	Design assessment required
Plane stresses	Impeller blades (centrifugal)	NRF
Static loadings	Mounting brackets	NRF
Manufacturing tolerances	Sliding coupling fits Shaft bearings	Check limits and fits
Assembly alignments	Coupling alignment	Check on design assembly alignments
3-D 'complex' stresses	Casing stresses Impeller blade fixings	NRF
Dynamic loadings	Shaft and key	Shaft and key
Fluid mechanics aspects	Volume air throughput	Data sheet 'throughput' check required only
Kinematic aspects	Location of frame mounting points	NRF

Important areas are shown shaded

NRF: Not relevant to the failure.

Fig. 6.8 shows the key size check.

Fig. 6.9 shows the assessment of limits and fits.

Figure 6.7   Step 1: The mechanical design parameters

- The use of *standard engineering data*. In this example BS 4500, reference (**3**), and BS 4235, reference (**4**), for limits and fits and manufacturing tolerances. Use standard data whenever possible – it saves effort.

The most important parts of these design assessment activities are the *conclusions*, shown shaded in Figs 6.8 and 6.9. Note how I have contracted them down to single unambiguous statements. These will inevitably be quoted in subsequent reports and minutes of meetings so they need to be clear and accurate. The overall message of this first stage of design assessment is as follows:

- no design *inconsistencies* have been found
- there is no evidence of inadequate design *limits*

but

- there are two possible contributing factors to the failure – the proximity of the end of the keyway to the change in section of the fan shaft, and the sharp profile and rough surface of the keyway itself.

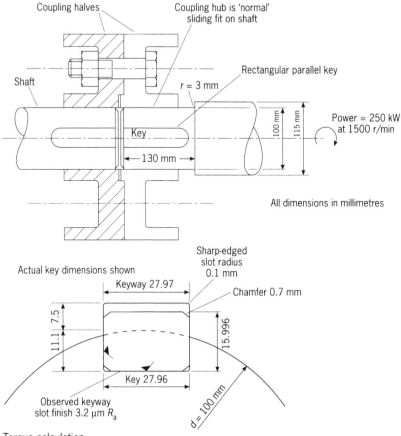

Torque calculation

$$\frac{T}{J} = \frac{\tau}{r} \quad \text{where} \quad J = \frac{\pi d^4}{32}$$

$$\tau = \frac{Tr \times 32}{\pi d^4} \quad \text{where} \quad T = \frac{60P}{2\pi N} = \frac{60 \times 250 \times 10^3}{3000\,\pi} = 1591.5 \text{ N m}$$

$$\tau = \frac{1591.5 \times (0.1 - 0.0111) \times 32}{\pi (0.078)^4} = 17.08 \text{ MN/m}^2$$

17.08 MN/m² << limiting τ of approx. 75 MN/m² (from material properties)

Key size calculation

Key area under shear = 0.028×0.13

$$\text{Shear stress} = \frac{F}{A} = \frac{1591.5}{0.028 \times 0.13 \times 0.05} = 8.75 \text{ MN/m}^2$$

8.75 MN/m² <<< limiting shear stress

CONCLUSION: BASIC STRESS CALCULATIONS OK
OBSERVATION: SHARP-EDGED KEYWAY SLOT AND 3.2 μm FINISH

Figure 6.8   The design assessment – keyway and key sizing

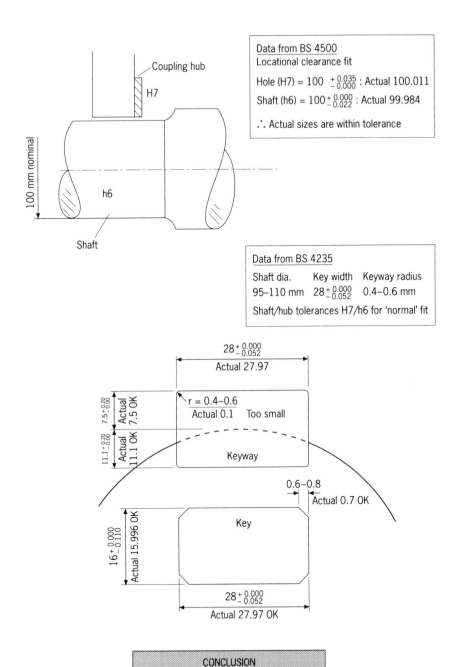

Figure 6.9 The design assessment – limits and fits

Note the apparent paradox here – it is stated that no major design inconsistencies have been found but the assessment has raised possible *contributory factors* to the failure. Findings like this are not uncommon in the design assessment part of a failure investigation. It is not really a paradox, rather a consequence of the complex nature of engineering design which means that, with the best intentions, design is not a precise science. There are always gaps and inconsistencies – if you look carefully enough.

This first step in the design assessment is not yet finished. It is little use raising the idea that the profile and location of the keyway end were contributory factors, and leaving it at that. If you do that, all you do is raise further open-ended questions – it is answers that are important, not further questions. Sadly, this is an all-too-common feature of many failure investigations and is a sure-fire way *not* to reach proper conclusions on causation. Try to follow this guideline:

If you raise technical questions – make sure you *answer* them as well.

In our example the 'answers' take the form of defining fully the technical aspects surrounding the positioning of a sharp edged keyway slot near a change in shaft section. It is necessary to quantify the effects of this configuration so that it is possible to make precise statements about the design, rather than just expressions of unspecified discontent. From basic engineering principles, the existence of slots and changes of section in rotating shafts has the effect of causing *stress concentrations*. These can multiply the torsional (shear) stresses by significant factors, encouraging local yielding and the initiation of cracks, leading to failure. Figure 6.10 advances the analysis, introducing standard reference data for stress concentrations in rotating shafts. Using such reference data is essential – it would be technically difficult, if not impossible, to develop such data from first principles in the time available. The conclusions of this further advancement of the design assessment are shown at the bottom of Fig. 6.10. Again, note how they make a *definitive* statement – one that can actually add something tangible to the investigation's subsequent conclusions on what caused the failure. This requires statements which are not only firm but also *quantitative*. Engineering design is a quantitative subject, so it is necessary to quote actual values of parameters such as stresses, stress concentrations, and the like. You will find that this helps to 'close out' the various steps of design assessment. It provides *answers*. You can see how the conclusions shown in Fig. 6.10 are used in the later stages of the failure investigation by looking forward to Fig. 9.13 in Chapter 9.

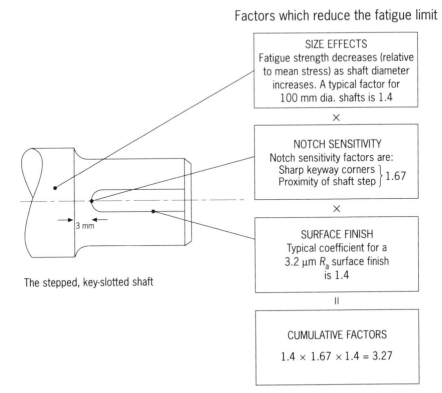

Factors which reduce the fatigue limit

SIZE EFFECTS
Fatigue strength decreases (relative to mean stress) as shaft diameter increases. A typical factor for 100 mm dia. shafts is 1.4

×

NOTCH SENSITIVITY
Notch sensitivity factors are:
Sharp keyway corners ⎱
Proximity of shaft step ⎰ 1.67

×

SURFACE FINISH
Typical coefficient for a 3.2 μm R_a surface finish is 1.4

=

CUMULATIVE FACTORS

$1.4 \times 1.67 \times 1.4 = 3.27$

3 mm

The stepped, key-slotted shaft

Hence the fatigue limit of the component is reduced by *more than two thirds* as a result of the design of this shaft configuration

Figure 6.10   The design assessment – reduction of the fatigue limit

Step 2: Check the power/torque capability
Power/torque capability is treated as a separate step because of its importance as a design parameter in almost any piece of equipment that has rotating parts. Figure 6.11 shows the basic design assessment calculation – note how these are based on 'steady state' dynamic conditions and then a factor applied to anticipate dynamic 'shock' loadings during driving, starting, stopping and transient conditions. This demonstrates an important principle of design assessment – that it is *dynamic* conditions that cause the greatest stresses and are therefore often instrumental in contributing to component failure. Components rarely fail as a result of steady state stresses unless they have a major design fault. The conclusions of Fig. 6.11 do not show any significant

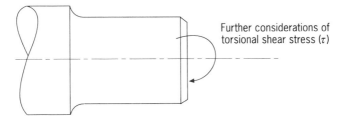

Further considerations of torsional shear stress (τ)

Torsional shear stress (τ): from Fig. 6.8 is estimated at ≈ 17 MN/m^2

Applying a factor of 4 to allow for dynamic shock–load conditions under starting/braking of the fan shaft

$17 \times 4 = 68$ MN/m^2 : much nearer the limiting value for the material

A review of possible bending due to bearing wear

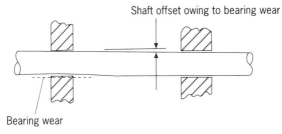

Shaft offset owing to bearing wear

Bearing wear

For a measured bearing wear (outside spec. limits) of 0.1 mm, the increase in bending stress σ_b caused by shaft offset is generally 'second order' compared to other stresses on the shaft (for this configuration). This can be calculated using standard theory.

> **CONCLUSION**
>
> Dynamic effects reduce the design factors of safety
> The observed bearing wear is unlikely to have contributed significantly to the shaft failure

Figure 6.11   The design assessment – further torsion and bending reviews

torsional strength problems; the design factors of safety are typical of those used in equipment designed for continuous rotation. Additional bending of the shaft due to the observed bearing wear is also evaluated in Fig. 6.11.

The other major parameter to be covered in step 2 of the design assessment is *unbalance and vibration*. For any component rotating at greater than about 100 r/min, residual unbalance of the rotating parts can be a contributory cause to many failures. The higher the speed, the greater the chance that unbalanced forces will be involved in some way. Unbalanced forces cause excessive vibration, setting up high stresses and fatigue conditions. Figure 6.12 shows the use of two of the most commonly-used technical standards covering unbalance: ISO 1940-1, reference (5), and ISO 10816-1, reference (6). These standards demonstrate one of the weaknesses, discussed earlier, of technical standards. They are indicators of sound technical practice, granted, but they are also documents of consensus. They are each the product of a committee – which says it all. This means that in your design assessment you must not expect them to give you all the answers. ISO 1940-1 is a typical example: it gives clear numerical limits on residual unbalance but the guidance as to exactly which limits suit which types of equipment is, without doubt, open to interpretation. Similarly, ISO 10816-1 gives 'classes' of vibration velocity acceptance criteria but is vague on which class is required for a particular piece of equipment. Note how in Fig. 6.12 the assessment shows that the fan rotor (the important part, as it has the largest radius of gyration) is found to be just 'on the limit' of the nearest recommended balance and vibration grades stated in the two standards. This is the place to use your engineering judgement. The question is in two parts:

- Is the documentated level of unbalance likely to cause excessive vibration of the fan rotor?
- If it did, could the vibration be the proximate cause of the observed shaft failure?

Don't confuse this with an either/or situation. Only if your 'engineering judgement' answer to both questions is 'yes' can you interpret the documented unbalance level as a true design error, to be taken into consideration when deciding causation. It can be tempting to weight your judgement *too* far towards discovering a design fault merely because you know that there *has* been a failure. Use your engineering experience to help your objectivity. This is another way of saying that maybe you shouldn't jump too quickly to conclusions. My view on Fig. 6.12 is that, taken alone, it *does not* provide enough real evidence for classing the residual unbalance of the fan rotor as the main 'cause' of the failure. We will see later how the operation assessment can add further to this picture – if there is clear evidence of even higher vibration levels it could develop the conclusion further, or even change it.

Suggested balance grades: from ISO 1940–1

Balance grade	Types of rotor (general examples)
G 1	Grinding machines, tape-recording equipment
G 2.5	Turbines, compressors, electric armatures
G 6.3	Pump impellers, fans, gears, machine tools
G 16	Cardan shafts, agricultural machinery
G 40	Car wheels, engine crankshafts
G 100	Complete engines for cars and trucks

Nearest designation to the case study radial fan

'Acceptance criteria': from ISO 10816–1

Zone 'D': Chance of equipment damage

Zone 'C': Usuitable for continuous long-term operation

Zone 'B': suitable for unrestricted long-term use

Zone 'A': for newly commissioned machines

Vibration velocity, V_{rms}

Frequency, f

Typical 'boundary limits': from ISO 10816–1

V_{rms}	Class I	Class II	Class III	Class IV
0.71	A	A	A	A
1.12	B	A	A	A
1.8	B	B	A	A
2.8	C	B	B	A
4.5	C	C	B	B
7.1	D	C	C	B
11.2	D	D	C	C
18	D	D	D	C

Measured result

Class suitability

Class I	Machines < 15 kW
Class II	Machines < 300 kW
Class III	Large machines with rigid foundations
Class IV	Large machines with 'soft' foundations

(Note how wide these classes are)

CONCLUSION
The measured vibration level of 7.1 mm/s is outside the apparent acceptable limit – but the ISO 1940–1 and ISO 10816–1 standards are not intended to be precise acceptance levels. Their text says they are 'for guidance only'

Figure 6.12   The design assessment – balance and vibration levels

Step 3: Check the wear resistance
Wear resistance is almost entirely a function of the materials of construction. The main factor controlling a wear regime is the relative resistance to wear and abrasion of adjacent moving components such as shafts/journal shells, piston rings/rings, etc. In practice a fair assessment of wear resistance can be made by simply testing a material's hardness (HV, HRB) value, normally given as part of the material's mechanical properties specification. As a 'rule of thumb', it is unusual for adjacent rotating parts to have a hardness difference of less than about 75 HB.

For the radial fan example the issue of wear resistance is perhaps not a major consideration in this particular post-failure assessment. Although some wear was identified in the roller bearing (see Fig. 6.6), inspection measurements indicate that it is sufficiently small to be classed as general 'wear and tear' (engineering judgement needed here again). There is no technical evidence to suggest that it is in any way part of, or contributed to, the shaft failure. In general, failures in proprietary pre-assembled bearings such as roller and ball bearing units are catastrophic rather than gradual. The bearing race tends to disintegrate within a short time, causing overheating and failure, so it is normally obvious whether or not an evaluation of wear resistance of the component parts needs to be a significant part of your design assessment.

Step 4: Check the temperature capability
The temperature capability of particular engineering designs manifests itself in three main forms:

- Resistance of component materials to *general process temperatures.* The most important case is for temperatures above about 390°C, when steels become susceptible to creep.
- Resistance of component materials to *local temperatures,* often caused by friction in rotating bearings. Local temperatures in highly loaded journal or thrust bearings can rise to 150°C, sufficient to cause distortion and melting of low temperature (mainly non-ferrous) bearing alloys. The heat can easily spread by conduction to gaskets, seals and other components critical to the operation of the piece of equipment.
- The effect of temperature on *design clearances.* Thermal expansions can reduce or increase design clearances to unacceptable levels if temperature design variations have not properly been considered, and allowed for, at the design stage. This type of problem is most common on precision rotating equipment, which relies on fine radial

and axial clearances on moving parts, and on equipment such as gas turbines and other highly rated prime movers where operating temperatures have large variations. Thermal expansions are difficult to anticipate during a design assessment because they operate in three dimensions and can only be analysed properly using computer finite element techniques or similar. It is sometimes possible to calculate single plane expansions (for instance the thermal expansion of a gearbox-to-coupling drive train) with a fair degree of accuracy but the significance of any conclusion will rest heavily on sets of assumptions that you will have to make to perform the analysis. Frankly, your best chance of identifying temperature-affected design clearances is *after the event*, by the nature of the damage or failure that has occurred. Thrust bearings sometimes fail like this, but it is always necessary to first eliminate other, more common, causes such as lubricating oil supply problems or simple overloading. The radial fan failure is a typical example – although there is evidence of wear in the shaft roller bearing, the application makes it most unlikely that this is anything to do with temperature-induced clearance problems.

During failure investigations, you should be careful not to place too much emphasis on looking for temperature-related design 'inconsistencies'. Temperature capability *is* a key design issue – it is also a well-known one, with a lot of research effort over the past fifty years targeted at the development of temperature-resistant materials. This means that you are unlikely to find glaring design inconsistencies with material *specification* – more likely any problem will be due to the material actually used not being the specified one. This is uncommon in mass-produced components but by no means rare in one-off items, fabrications and castings.

Step 5: Check the resistance to corrosion
Corrosion resistance is another of the criteria by which the materials of construction of a component are selected. Stainless steels, high nickel alloys and non-ferrous materials all play their part in minimizing corrosion in corrosive process conditions and, again, a lot of development has been done. In spite of this, corrosion resistance is far from being a perfect science, for two main reasons:

• Corrosion is unpredictable. Even with identical process conditions, different components made of the same material can corrode at different rates. The same applies to typical corrosion rates quoted in

learned technical papers and data sources – they are rarely repro-
ducible in use.

- Process conditions always vary. Slight, unpredictable variations in
process parameters; temperatures, fluid velocities, etc. can cause
large differences in the corrosion performance of components'
materials. This adds to the overall picture of unpredictability.

Don't even *start* to get involved in a detailed study of corrosion rates
during a failure investigation design assessment – unless you are a
specialist in this field it is a technical cul-de-sac. For investigations
where corrosion seems as if it may have played a part in causation, it is
best to stick to a basic technical assessment of the corrosion 'suitability'
of the materials used for the failed components. Limit the assessment to:

- Basic electrochemical series assessment (from published data)
- Resistance to strong acid/alkali conditions and well-known corrosive
compounds such as metal chlorides.

You are only looking for obvious design inconsistencies such as
unprotected low carbon steel or cast iron used in sea water. This
means serious, obvious failures – don't be drawn into assessing whether
a design *encourages* corrosion because of its physical configuration, flow
velocities or other detailed design parameters. You will soon get out of
your technical depth.

Step 6: Check compliance with technical standards
This is a tricky area. Published technical standards cover many areas of
engineering practice and form an important 'skeleton' of reference
information. They are documents of sound engineering practice, undoubtedly, but must be viewed within the context in which they are
written. Here is an outline of the properties of published technical
standards:

- Published technical standards demonstrate good technical practice

and

- are (justifiably) accepted as authoritative by most technical parties

BUT

- rarely cover all technical areas of a piece of equipment

and

- are documents of *consensus* – written by committees – and so can sometimes be ambiguous, indecisive or even inconclusive.

These last two points summarize neatly the real weakness of those technical standards dealing with the design and manufacture of *equipment items* (rather than smaller machine element components such as screw threads, pins or clips). The message is that you can, and should, be happy to rely on published technical standards but you must bear in mind their *limitations*.

How does this affect your post-failure design assessment? The main implication is that, as with the general philosophy of design assessment suggested earlier, you should look for *non-compliance* rather than compliance. This way you will not suffer from many of the inadequacies of technical standards but you will be able to take advantage of their strong points. Major non-compliances will become strong findings of your design assessment activity. Be careful also not to *underestimate* the uses of technical standards – whatever their limitations, they are still authoritative documents, widely accepted, and undoubted examples of good engineering design and manufacturing practice. They really are very difficult to argue against. Looking back at the general methodology of design assessment shown in Figs 6.1 and 6.2 you can see how technical standards can impinge upon fitness-for-purpose issues in all six assessment 'steps', which is why they are such a key part of the assessment process.

Figure 6.13 shows the situation regarding standards compliance for the radial fan failure case. Note how the relevant standards are first *identified* for each technical area under consideration and then compliance points referenced individually. This is the only thorough way to do it – it is not sufficient simply to make generalized statements that the equipment 'complies with all standards', without showing that you have looked carefully at the important technical detail. There is a sound reason behind this – when you reach the later 'causation' stages of the investigation you may be faced with the comments of specialists, and you will soon become exposed if your design assessment is in any way weak or incomplete.

Recording the results
It is of little use putting a lot of effort into a detailed design assessment and then not recording the results properly. *Properly* means that other people, who may not be experienced in engineering, can understand them easily. There are two points to avoid:

FAILURE: RADIAL FAN SHAFT

Preliminary details

Failed shaft – some bearing wear – marginal balance – loose coupling bolt – corrosion on external parts.

Component	Standard	Compliance point
General fan design	BS 848 (1980)	Testing and performance
Dynamic balance rotor	ISO 1940-1 and ISO 10816-1 (1996)	Grade G6.3 (equivalent to 6.3 mm/sec vibration velocity)
Shaft material specification	BS 970 (1996) (ref 7)	Mechanical properties
Shaft key size	BS 4235 (1972)	Key and keyway dimensional and profile tolerances
Coupling bolts	BS EN 3052	Material compliance
Limits and fits	BS 4500 (1985)	H7/h6 sliding fit for coupling hub and shaft

---

Conclusion

Non-compliances have been found: ISO 1940-1 (marginal non-compliance on grade) and BS 4235 (significant non-compliance on profile of keyway).

Figure 6.13   Design assessment – standards compliance check

- The purpose of your design assessment results is not to help you obtain a doctorate, so don't try.
- Don't encourage indecision by being guarded and inconclusive – a design assessment that only gives heavily qualified contingent answers is of little real value.

The best way to guard against these all-too-common mistakes is to use a simple pro-forma summary report. This records the extent and scope of the assessment (Fig. 6.2) but keeps the answers simple. It can be used by insurers, managers, even lawyers, as a clear record of the technical findings. Done properly, it will be of more use than any amount of voluminous reports and technical data.

Figure 6.14 shows the overall design assessment summary report for the radial fan. Note the form of wording used – it makes careful but not too-generalized comments about what was *not found*. This demonstrates an important point about the way to report the results of design assessments. Figure 6.14 is not saying that the design of certain fan components was definitely unrelated to the failure, all it is doing is eliminating those areas that were looked at during the design assessment – there are many other aspects of design that could somehow be involved. So, here is the big question:

Area of assessment	Result
Mechanical design parameters	Keyway profile too sharp and has $3.2\,\mu m\,R_a$ surface finish
Power/torque capability	No major inconsistencies found
Resistance to wear mechanisms	No major inconsistencies found
Temperature capability	No major inconsistencies found
Resistance to corrosion	No major inconsistencies found
Compliance with technical standards	Non compliance with ISO 1940-1 and BS 4235

Conclusions

The design assessment showed no significant design deficiencies which, taken alone, could be considered the single cause of failure of the fan. The non-compliant keyway profile will result in increased stress concentrations and lower the resistance of the shaft to fatigue conditions.

Figure 6.14   Design assessment summary for the radial fan

• If the fan design is 'satisfactory', why did the shaft fail?

Teasing out the answer to this, getting at the real issue of causation, is one of the fundamental tasks of the explanations in this book – we will see the answers start to develop in the next chapter. Try to accept, for the moment, the idea of the design assessment as an exercise in *elimination* – as one of the several elements of effective failure investigation. It is an essential element, and it needs to be done thoroughly, often with some precision – but it is still only *part* of the story.

The operation assessment

The purpose of the operation assessment is to find out whether the way in which a piece of equipment was operated had anything to do with its failure. Operational factors often take the 'instant blame' when mechanical failures occur, whether justified or not. They are often stated as being the cause of failure when a failure investigation has commercial implications – because it is contractually convenient for manufacturers to claim that their equipment has only failed because it was operated 'outside its design conditions'. This encourages a degree of wishful thinking by manufacturers when they are investigating failures of their own equipment (which they often do). Such situations highlight the need for a clear and objective assessment of operational factors to be part of any effective failure investigation.

In a similar way to the design assessment, the operation assessment is primarily an exercise in *elimination*. The objective is to eliminate, as far as possible, operational factors as contributing to the cause of the failure. Expect the methodology to be a little different to that of the design assessment – it is still fundamentally an information-gathering exercise but simpler, with only two main parts. These are:

- *Identifying operational limits* – primarily a desk-based exercise consisting of deciding the form and limits of various operational 'envelopes' for the equipment concerned.
- *Clarifying the operational history* – this is more a site activity involving the review of operation logs and records and then validating their content via discussions with operations staff. The techniques of interviewing introduced in Chapter 4 play an important part.

The difficulty of both of these parts of the assessment is that they involve looking into the past. The operational limits are a combination of design and operational factors, both historical, and operational history, however recent, can soon become shrouded in secrecy once it is known that a failure investigation is underway. Both parts require an active and searching approach if they are to produce any meaningful results. This is the place for the type of active offence introduced in Part I of this book.

Identifying operational limits
The inventory of relevant operational limits depends on the type of equipment that is being investigated. Large items of mechanical equipment which have many moving parts will obviously have a greater number of operational constraints than, for example, a small pressure vessel, which is of simple construction and does not move. Luckily, areas of technical commonality between many types of equipment mean that a pattern does emerge. As a *very general* guideline, for rotating mechanical machinery, the following operational 'envelopes' are the important ones to look at:

- operational hours or cycles
- speeds
- temperatures
- reliance on auxiliary systems (mainly lubricating oil and cooling water/air)
- mechanical loadings.

- One life-limiting feature of the fan's design is the effect of high-cycle fatigue of the shaft (bending and/or torsional stresses)

- A basic 'reference characteristic' is shown for the shaft steel grade approximating to BS 970 type 080M40 (normalized)

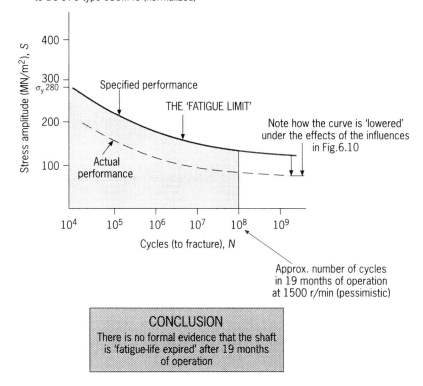

Figure 6.15(a)   The operation assessment – looking at fatigue 'lifetime'

You can think of these as the lowest common denominators of successful operation – if they are correct, the equipment will work satisfactorily. They are in no particular order, all contribute about equally. The principle is first to identify the acceptable 'shape' and limits of the operational envelopes and second, using the results, to find out if the equipment has ever been operated outside them. If it has, this may be relevant to (or even be the proximate cause of) the failure.

The best way to demonstrate this approach is by using the case study example. Figures 6.15(a) to (d) show the four key operational 'envelopes' for the radial fan. The way in which these have been compiled is important – they are a combination of manufacturers' specification data and more general technical information based on conventional engineering experience and judgement. This is why a wide technical

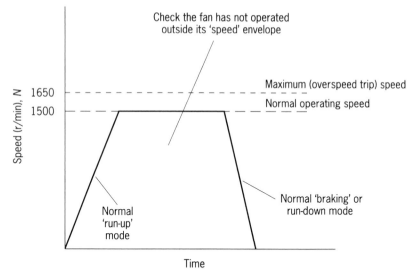

Figure 6.15(b)   The operation assessment – check the operating speeds

background is so important for failure investigators – the necessary background information is invariably written down somewhere, but you have to know where to look. As a guideline, you can expect that *about half* of the key information in the operational envelopes needs to come from your experience, the rest you can source from the various specifications and data-sheets for the failed equipment. I have tried to show you in Fig. 6.15 which is which.

Operational lifetime
Figure 6.15(a) shows the situation as assessed regarding operating 'lifetime factors'. For a fan, unlike for instance a pressure vessel, the components have no formal design lifetime as such. Practically, it is the fatigue life of the cyclically-stressed components that forms the life-limiting factor. This type of information (and the typical S–N characteristic shown in the figure) is unlikely to be provided in the manufacturer's specification so it has to come from more general data sources, or experience. In this case a suitable ISO technical standard was used to source the S–N curve.

Running speeds
Running speeds are important for rotating equipment because they have an influence on the stresses imposed on both rotating and static

Check the configuration of the fan's auxiliary systems

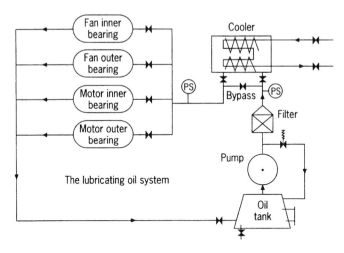

The lubricating oil system

The fan commissioning network – check for compliance

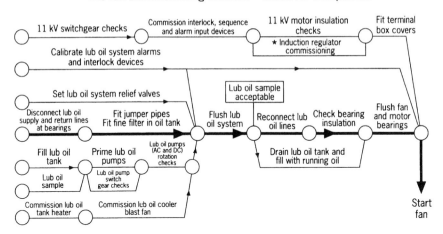

Figure 6.15(c)   The operation assessment – check of auxiliary systems

components. For the fan, the assessment is straightforward, the operational 'envelope' is the range of speeds up to the maximum synchronous speed of 1500 r/min (Fig. 6.15(b)). For variable speed equipment such as engines or turbines, the situation would be more complicated. Note how the figure could show the speed limits for different guide vane settings – there might be stress criteria which place a limit on the acceptable speed in each configuration. Note also the torsional shear stress calculations in Fig. 6.8 which shows how

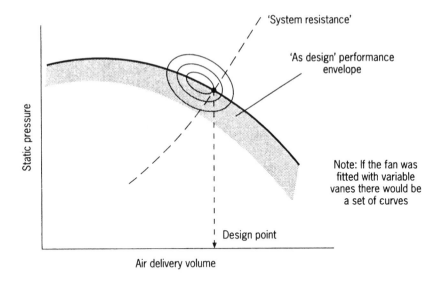

Figure 6.15(d)   The operation assessment – performance 'loading'

general technical theory could be used to do a rough check on stresses resulting from the rated, or 'trip level' rotational speed. An assessment of absorbed power or torque could also be done for the fan, adding to the detail of the operational envelope.

Reliance on auxiliary systems

This relates to the fact that all equipment items rely, for correct operation, on the satisfactory performance of auxiliary systems supplying lubricating oil (LO), cooling air, etc. Again the example of the fan is straightforward – Fig. 6.15(c) shows the installed LO system (from the manufacturer's specification documents). Here the operational 'envelope' is the situation whereby all the components of the LO system are in the correct configuration and have the correct status (open, closed, etc.). A further interpretation of the operational envelope would be the requirement that the network of fan commissioning activities shown was completed correctly each time the fan was used. Try to comprehend these different *viewpoints* on the idea of the 'operational envelopes' – once you have mastered the idea, you will find it a useful and streamlined way to structure your investigation into the causes of failure. It should also make your reports more focused, and easier to understand.

Mechanical loadings
For fans, the limits on mechanical loadings are closely related to the fluid *performance* of the unit. This is common to other types of rotating equipment, each of which has a performance envelope within which it is designed to operate. Figure 6.15(d) is the manufacturer's curve for the specified fan showing clearly the operational envelope. There are often some unknown elements in performance envelopes like this – here it is the 'system resistance', but it varies between equipment types. Expect variety in the way envelopes are expressed – try to see through to the *principles* involved.

Clarifying the operational history
Having ascertained the limits of the operational envelopes, the next step is to find out about the operational history. This is one of the early steps in the analysis of what actually *happened* to the equipment. It is mainly a site activity, involving discussion with operations staff, so it is an exercise in asking and listening, using the principles outlined in Part I of this book. The technique itself is best organized into timescales, to keep the activities in order. Figure 6.16 shows the basic structure of three stages. These stages can be fairly streamlined as the purpose at this stage is only assessment – the analysis stage comes later.

Operation before the failure event
This is the broadest part of the assessment. Your discussions should cover the operation period before the failure occurred with one main objective: to confirm whether the equipment was really being operated within its operational envelopes. Keep clear of discussing the failure, for the moment; operations staff will be more comfortable with this approach. Try to address each operational envelope separately – if they do become 'mixed up' in the discussions then separate them in your own mind, before coming to conclusions or making notes. A further area to investigate is *external factors*; whether there were any outside happenings or influences on the equipment (mainly from its process system) that were having an effect on how it was operating. Items of evidence such as log sheets, documentation or records of trip settings are useful in giving you a level of confidence about what you are being told.

The failure event itself
Expect colourful explanations about the failure and what everyone thinks caused it – you will rarely be disappointed by their variety. Unfortunately, you *must* listen to them all because they will each have

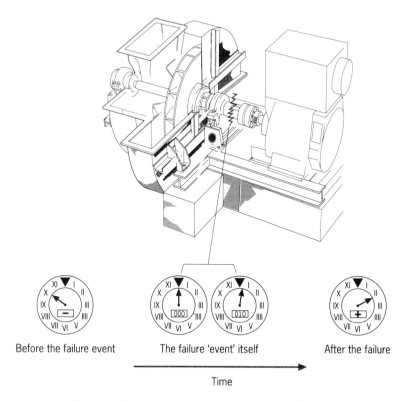

Figure 6.16   The three stages of operational history

something to add to your overall understanding of the failure. This part of the investigation is about picking out and marshalling the multitude of small technical *facts* that are revealed during the discussions. You need to tease these facts out from the confusion (each will be surrounded by a wall of opinion) then record them carefully in writing for interpretation at a later stage. This is an important procedural point; it is not wise to form instant opinions before completing the analysis. It is worth repeating – keep this as a fact-gathering phase.

After the failure
This refers to happenings in the few seconds or minutes *immediately after* the failure event. Failures of engineering components frequently affect other parts of the equipment, often to the extent of causing actual consequential damage. Because of the uncertainties that inevitably follow a major failure it can sometimes be difficult to get a detailed understanding of precisely what happened. A wide-ranging investiga-

tion gives the best chance – the following areas can be useful 'lead-in' points to the discussions:

- Which alarms and trips operated?
- Was there any operator/manual intervention made (opening or closing valves, emergency shutdown, etc.)?
- Were any physical changes made, that changed the equipment in any way from the pre-failure running configuration?
- What kind of records were taken of the state in which things were found?

All of these areas can contain important factual clues. They will often be partially hidden in a lot of peripheral information so it is necessary to tease them out. This chapter has covered in some detail the procedure of the design and operation assessments. These are an essential part of the failure investigation and are also good examples of the principles and *difficulties* of an assessment. Facts rarely present themselves in the form that you would like – they have to be teased out, from a mass of other technical information and opinion. You can't expect this to be easy, but it is the quality of these assessments that governs whether you can properly 'break open' a failure investigation or whether it remains closed, leaving you searching for convincing technical answers. A thorough and vigorous approach is what is needed to give yourself the best chances of success.

KEY POINT SUMMARY: DESIGN AND OPERATION ASSESSMENT

1. Objectives

The design and operation assessments are exercises in elimination. They are more likely to tell you what did *not* cause the failure than what did.

2. Methodology: design assessment

There is a well proven methodology for doing a design assessment (see Fig. 6.1). Some parts of the assessment are naturally harder than others. The steps are:

- mechanical design parameters
- power/torque capability
- wear resistance
- temperature capability
- resistance to corrosion
- compliance with technical standards.

3. The operations assessment

There are two main parts to this:

- Identifying operational limits of a piece of equipment.
- Clarifying its operational history.

Expect the full details of operational history to be difficult to find. A good assessment involves asking lots of (difficult) questions about what happened before and during a failure incident.

Chapter 7

Categorization

Categorization -v- causation: the difference

This is mainly a question of depth. The depth or *resolution* to which a technical problem is addressed is important – it is not always easy to get it right. Fortunately, it is just about 100 percent learnable. A useful first step is to recognize the split between categorization and causation:

- *Categorization* is the technical step of recognizing and defining the nature of a failure mechanism.
- *Causation* is more about *describing* the failure in a way which is meaningful in both technical and various other terms. The activity has a technical root but is not, itself, entirely technical – there are 'procedural' parts to it, and aspects of convention.

It is difficult to over-emphasize the importance of this split. It is an important part of clear thinking in the later stages of a failure investigation. This chapter covers categorization, leading on to the description of the causation stages in Chapter 8.

Categorization

Categorization is the natural follow-on from the activities of the failure investigation inspection visit described in Chapter 5. It is a pro-active activity – you can think of it as a broad-brush technical review of the failure; complete in its technical detail, but without many of the finer presentational points that are needed for it to qualify as a proper conclusion on what caused the failure.

Is categorization difficult?

Given the variety of engineering equipment that exists (and can fail) there are potentially a large number of different failure types and mechanisms, so you cannot always expect it to be straightforward.

Fortunately, engineering failures tend to fall into a manageable number of categories which makes them easier to handle. So:

Categorization becomes easier if you know what to expect.

Figure 7.1 contains basic and well-proven data about engineering failure categorization. The strength of this data is in its applicability to failure cases that you will meet. Note how the failure categories relate to components (and processes) that are common to many different pieces of engineering equipment. This data is not perfect, or exhaustive, but it fits well with the majority of mechanical equipment failures. Over the course of perhaps twenty or thirty investigation cases, the findings should sort themselves out broadly as shown in the amalgamated data of Fig. 7.1. Three points stand out:

- It is *stressed components* that fail.
- Of these stressed components, failures of *welds and shafts* account for more than half of all failures.
- About 80 percent of failures are caused by some sort of *fatigue*. Static fracture, corrosion and wear are much less common.

Remember that these are essentially technical categories, comprising well-defined and precise definitions of the mechanisms of engineering failure. It can be misleading to use them too loosely – too many failure investigations lose direction because of the over-use of loose technical definitions. It is necessary to look closely at the real engineering meaning of these failure categories. First, though, let us consider a slightly controversial background point – the role of metallurgy, or (its more general description) materials science.

Paradox: the role of materials science

The argument looks like this:

- Materials science (i.e. metallurgy) is a robust theoretical and practical engineering discipline

but

- Its technical strength relies on its links with empirical results – unlike mathematics it is not fully predictive.

This leads to two further points about the practical way materials science fits into failure investigation:

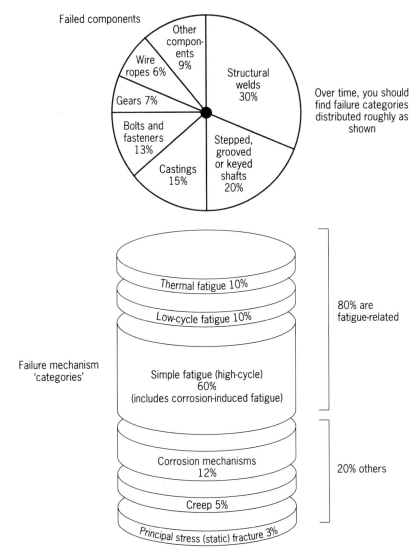

Figure 7.1   Failure categorization – what to expect

- You need metallurgical principles and knowledge to help 'solve' a failure investigation

    but

- they will rarely give you the complete explanation about what happened.

I have found these four statements to be true. I also maintain (although this is perhaps moving off the subject) that *only* these four statements are true about materials science – and that just about everything else is open to contradiction, interpretation, differences of opinion, and technical disagreement. Metallurgists, like economists, biologists, doctors, and accountants, will always disagree, once the discussion reaches a certain technical level. This is understandable – but it *won't help your failure investigation*. This means that from the diagnostic stages of a failure investigation onwards, you are faced with the realities of a discipline which has high complexity – you can think of it as having the capability of producing conclusions which are not conclusions, or which are unstable. You should be able to see evidence of this in the following sections of this chapter as we dissect each technical failure category. Materials science and metallurgy have their part to play – but they need to be controlled, if a failure investigation is to keep its direction.

Failure categories

With these conclusions about the role of metallurgy in failure investigations firmly established we can now look at the main failure categories and some typical ways that they manifest themselves.

Failure due to principal stresses

These are the simplest types of failure. The component fails owing to principal stresses, either in a three dimensional 'tri-axial' field (σx, σy, σz), or in the simplest cases, straightforward uni-axial or bi-axial stress. Principal stress failures can occur in components which are static, or which are moving (dynamic) in their normal operating mode. The important point is that they are operating within their design limits and not subjected to any *unexpected* imposed vibration, shock loading, temperature excursions or other non-design conditions. Unlike some other failure categories such as fatigue or creep, true principal stress failures have no specific characteristics or visual appearance. For example, both ductile and brittle fractures (which have vastly different appearances) can be defined, albeit under different conditions, as principal stress failures.

What causes principal stress failures? There are two basic causes: design *inadequacy* and operational *overload*. It is a basic design criterion that an engineering component should be designed so that it can withstand the principal stresses to which it is subject in its expected operational service. If it is not, or the component is overloaded in some way in service, then it may fail by a mechanism of excessive principal stress. This of course is the extreme case – if the design inadequacy or overload is less pronounced, then the component may deform or wear more quickly than normal, but will not fail immediately. This concept of timescale is a key part of categorizing failures – a regime of excessive principal stress that does not cause immediate failure can easily 'be responsible' for a failure that occurs later, perhaps by a different failure mechanism. The failure categorization will have *changed*. So as a general point:

When a failure occurs is always relevant when categorizing failures.

The main types of principal stress failure fit in well with the standard theoretical cases of static and dynamic loading found in any textbook. These are, by necessity, highly simplified and so need some inter-pretation to apply them to real engineering equipment. Figure 7.2 shows some common examples, applied to our case study example. We can look at each in turn, in a little more detail.

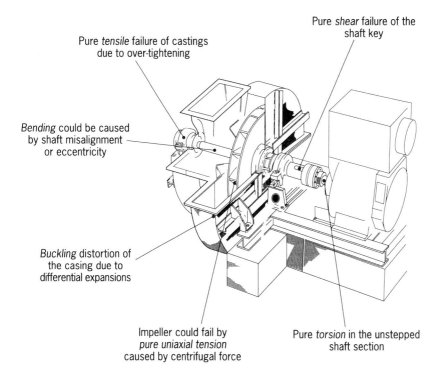

Pure *shear* failure of the shaft key

Pure *tensile* failure of castings due to over-tightening

Bending could be caused by shaft misalignment or eccentricity

Buckling distortion of the casing due to differential expansions

Impeller could fail by *pure uniaxial tension* caused by centrifugal force

Pure *torsion* in the unstepped shaft section

Figure 7.2   Possible failures due to principal stresses

Pure tension

If one of the fan's impeller blades was to break, with the fracture starting well towards the centre of the parent metal of the blade (i.e. away from any small cast radius or weld) the cause of the failure could be due to almost pure tensile stress. For the design shown, the centrifugal accelerations imposed on the blades during rotation produce a tensile stress which is predominantly uni-axial in the radial direction. Practically, such a failure would be quite rare unless the blades were unsupported as in, for example, a steam or gas turbine rotor. Complex engineering equipment components actually have very few areas that experience only uni-axial tensile stress – perhaps the most common ones are cast items such as bearing housings, static plinths and supports. Cast components can have poor tensile strength – they will fail by a principal tensile stress mechanism (normally bi-axial or tri-axial) if misaligned or over-tightened. Apart from these examples, failure by principal tensile stress is rare. There are not that many basic design mistakes. Most tensile fractures that do occur have ductile characteristics. They are generally caused by the slow application of load, and the region around the fracture face exhibits the classical features of 'necking'.

Pure shear

Shear stress tends to be a complement to tensile stress in engineering components. Traditional theory covers the way in which a principal tensile load on a test specimen results not only in tensile stresses parallel to the load but also in shear stresses which reach a maximum on planes inclined at 45 degrees to the principal tensile stress direction. In practice this means that a shear failure can occur in almost any plane in complex-shaped engineering components, particularly castings. In such a failure the effects of shear will be similar to those of a principal tensile stress failure and the precise theoretical explanation of which type of stress caused the failure becomes rather academic. Most shear failures are actually preceded by some bending.

A pure form of shear failure is shown in Fig. 7.2. If the shaft coupling drive key was undersized then it could fail, under normal operating conditions, by a mechanism of pure shear. This would occur in a direction tangential to the shaft surface. Practically, failures like this are quite rare – a good designer would incorporate a factor of safety of 7 to 10 in the dimensions of the key and keyway. The key could still fail (they do), but this would most likely be influenced by other failure mechanisms, rather than only pure shear. Other components which can fail by pure shear are gear teeth, fasteners such as bolts or pins, and badly configured welds. The fracture faces are generally flat, with no evidence of directional features.

Pure bending

Pure bending stresses caused by the bending of beams and other standard members are well covered in engineering handbooks and textbooks. There are basically three types of bending:

- simple bending of straight members (including beams)
- bending of curved beams
- complex bending.

By far the most relevant to engineering failure investigation is the third type: the large family of cases which can be described as *complex* bending. Bending may be complex owing to the indeterminate nature of either the material section or the bending load regime, or both. This fits well with mechanical engineering practice – few equipment components subject to bending are of simple plane or prismatic section, making it necessary to find out the section modulus before any meaningful calculation of bending stress can be made. The situation is easier if the failed component is of simple configuration: chain links, shackles, ropes or bars, and structural steelwork items such as columns and I-beams are good examples. Bending doesn't actually cause the failure – the final break is normally the result of a ductile or brittle fracture mechanism.

Figure 7.2 shows one easily identifiable pure bending failure mechanism that could happen to our case-study fan. It is feasible that misalignment of the shaft bearings could cause sufficient principal bending stresses on the shaft section to cause failure. There would also be torsional stresses in the shaft, due to its rotation, but it would be the bending effects of the assembly misalignment that would have the greater effect. Failures due to pure bending are slightly more common than those of simple tension or shear but still rare – equipment and structural design normally incorporate substantial factors of safety to protect against it.

Buckling

Buckling is a compression effect, caused by long and thin or otherwise flexible members (struts) being subject to too much compressive load. This causes them to bend excessively ('buckling') and hence lose their ability to act as a strut. Depending on the load, they may deform plastically and then break. The situation is made worse if the load is eccentric from the axis of the strut. The most common place to find buckling failures is in components made of large unsupported areas of steel plate which are subject to temperature variations. Stresses caused by constrained thermal expansions can be very high, resulting in buckling. Hollow cylinders subject to external pressure can also buckle if incorrectly designed. The theoretical case is similar to the Rankine formulae used for long struts and designs normally incorporate safety factors of at least five to guard against uncertainties in the theoretical calculations.

The example shown in Fig. 7.2 is a typical buckling failure. The sheet steel sections making up the fan casing have buckled due to constrained thermal expansion when the casing heats up to operating temperature. In extreme cases the buckling could distort the casing sufficiently to contact the impeller, possibly causing further damage. This is a good example of how failure mechanisms can have *consequential* effects – starting off other, different, failure mechanisms.

Pure torsion (twisting)
Torsion is a common 'principal stress' occurrence in engineering components. In its basic form, it produces simple shear stresses which reach a maximum value on a plane at 90 degrees to the component's axis. Theoretical calculations are easy for solid cylinder or hollow cylinder sections but become more complicated for non-circular sections. The torsion/shear stress relationships become progressively more complex as the geometry of a component departs from the simple concentric form. Torsional shear stresses can be set up not only in the more obvious rotating components but also in normally static members which are subject to twist. This adds to the complexity of the loading regime on even a simple engineering component – shear stresses resulting from pure shear, torsion, bending, and pure tension can add, or subtract, together to form a very difficult picture.

Failure by *pure* torsion, as for the other categories of principal stress, is not very common. The theory is well known and large factors of safety are employed to deal with uncertainties. Of those components that do fail in this way, many are static cast pieces that have been subject to unexpected twisting. Even then, pure torsion failure is rare because there are nearly always other factors such as stress concentrations or impact loadings involved. Figure 7.2 shows a classical, but normally unexpected, case – a pure torsion (ductile) failure of an unstepped prismatic section of the fan drive shaft.

To summarize, true principal stress failure in properly designed mechanical equipment is not just infrequent: it is, frankly, rare. Principal stress plays a part in the progression of a failure but it is rarely the initiator. Initiation (remember the concept of timescale?) is the important point when you are trying to determine and describe causation. So:

Don't expect to find many failures where excessive principal stresses were the 'main' cause.

So what does act as the initiator of failures? Sadly, there is no single answer – it would be nice if there were. There are several possibilities – we will look at these in, very broadly, ascending order of the frequency with which you should find them across a range of failure investigations.

Wear mechanisms
All engineering components experience wear to some extent but particularly those parts (loosely known as machine *elements*) that rotate, slide, or otherwise make contact with another element. Gear teeth, piston rings, and bearings of all types are the usual examples. The traditional view of wear is that of an even, progressive mechanism – the components wearing gradually in a more or less predictable way and producing no surprises. While this may not be exactly proven in practice, one point is beyond doubt:

Wear is always *time dependent.*

In addition, the time-dependency of wear is better understood than it is for some of the other possible failure mechanism 'categories'. Roller bearing design is a good example – bearing lives are quantifiable to the point where specific bearing designs are given life-classification numbers, predicting the number of hours of loaded operation before the roller and race components become worn beyond serviceable limits. Because of this degree of predictability, failures are rarely caused exclusively by wear, but it is often a contributory factor. There are two main possible types of wear – having in common the fact that they are the result of a *mechanical* process. We can look at these in relation to the way that they could occur in our fan case-study (see Fig. 7.3).

Abrasive wear
The most likely places for abrasive wear are:

- In the journal bearings (scoring), due to oil contamination.
- On the impeller surfaces (erosion), due to abrasive particles in the air.

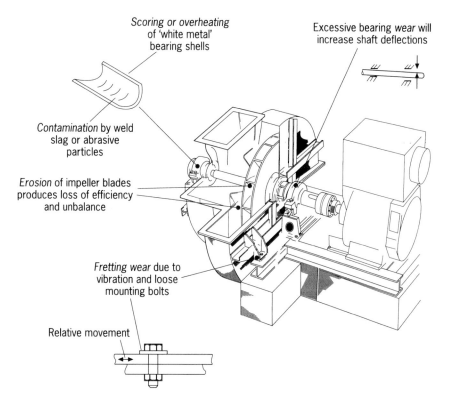

Figure 7.3   Some possible wear-related failures

The first of these, bearing wear due to oil contamination, is particularly common in rotating mechanical equipment. The most likely contaminant is welding slag or grinding debris from the inside surfaces of welded pipe joints. Both of the examples help indicate an important common factor of wear-related failures – they are mostly caused by a level of contamination which is outside the design expectation. This is a common occurrence with pumps, compressors and process equipment which have to resist an abrasive environment. The reason is not that design parameters are poorly understood, but that equipment designers have to deal with the inherent unpredictability of operating conditions; defining the abrasiveness of river water, for example, is difficult. Cost also plays a part – designers are always under pressure to specify the cheapest material that will (just) do the job for which it is intended. This all influences what you can expect to find during the design and operation assessment set out in Chapter 6 – don't expect abrasion wear resistance to be objectively and neatly defined (because it won't be). The bearing failure in Fig. 7.3, in contrast, is not related to any design-induced uncertainty. This is a pure manufacturing fault – the result of not removing the scale or contamination from the oil system before commissioning the equipment.

Fretting wear
Fretting is a wear regime caused by two surfaces in contact with each other being repeatedly rubbed together, often as a result of externally induced vibration. Bolted joints, castings, housings and mountings of all types can suffer. It can cause failure, but generally only over a long timescale. As with abrasive wear, the causes are predominantly ex-ternal – it is the cumulative effect of vibration and insufficient tightening during assembly that starts the process. There are some important messages in these two simple examples. While wear is undeniably an identifiable category of failure, capable of robust technical definition and explanation, its significance is contingent on two things:

- Whether there has been any *mis-selection of materials*. Bad material choice will exaggerate wear.
- Was the wear aided by any outside *influences*? The contamination (or starvation) of lubricating oil and airborne abrasive particles scenar-ios in Fig. 7.3 are typical examples of this.

If you exclude these two contingencies, then failure due to wear is uncommon. If either of these contingencies do occur then the resultant failures may be described as 'wear-related', but wear is not the proximate cause. It is the material choice or outside influence that provided the 'condition' for failure to occur – the wear itself is consequential.

Corrosion mechanisms
About 10 to 12 percent (see Fig. 7.1) of engineering failure incidents are directly due to surface corrosion mechanisms. Corrosion is responsible for limiting the useful lifetime of equipment in many other cases, but the equipment or component is generally replaced before a failure event occurs. Like wear, corrosion is a time-dependent mechanism – it has a gradual deteriorating effect over time, rather than causing catastrophic failure 'events' in the early life of a component. It is also analogous to wear in that it can proceed either with, or without, help from outside influences.

Corrosion can lead to failure in two ways:

- By reducing the *strength properties* of a material

- By reducing the *section thickness* of a component (Fig. 7.4 shows a good example)

Either (or both) of these reduce the strength of a component to the point where it cannot support its design loads, leading to failure. Ultimately, as the component's strength reduces, the failure will manifest itself as one of the principal stress failures, but the mechanism of corrosion is usually so well identifiable by this stage that it remains inextricably linked with the failure event itself. Hence corrosion mechanisms *are* considered to be a 'proximate' cause of failure – unlike wear, which retains a more tenuous relationship with the actual failure event. A gamut of formal terminology exists to describe the many different forms of corrosion, mainly of ferrous materials. To maintain a clear focus on failure categorization (the purpose of this chapter) it is wise not to become too involved in this. The effect and, as we will see in the next chapter, the *cause* of corrosion are two key issues, not an elaborate technical description of the chemistry of the corrosion process itself. Unless you become involved in very specialized 'metallurgical' failures the following categories should be about all you need to categorize corrosion-related failures.

Chemical attack corrosion
This is caused by aggressive attack of the surface of a metal by strong acidic or alkaline compounds such as chlorides and other salts. It can produce a uniform corrosion rate across a surface but more likely results in localized corrosion pitting. Temperature is an important

Figure 7.4   A well-corroded boiler stay (Courtesy: Royal and Sun Alliance Engineering)

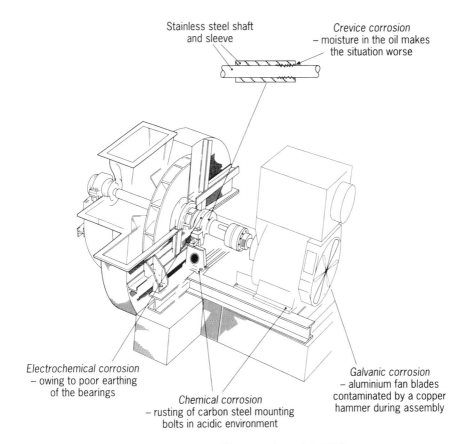

Stainless steel shaft
and sleeve

Crevice corrosion
– moisture in the oil makes
the situation worse

Electrochemical corrosion
– owing to poor earthing
of the bearings

Chemical corrosion
– rusting of carbon steel mounting
bolts in acidic environment

Galvanic corrosion
– aluminium fan blades
contaminated by a copper
hammer during assembly

Figure 7.5   Some possible corrosion-related failures

factor – the rate of corrosion nearly always increases with temperature. While the basic mechanisms of chemical corrosion are well known – you can find them in engineering handbooks – you may find that it is still a far-from-perfect science. The resistance of even well-known stainless steels such as 304 and 316 to chemical corrosion is still subjective, and carries with it a surprising amount of unpredictability. The same goes for specialist cast irons such as Ni-resist and other alloys developed specifically to resist chemical corrosion. For any individual design it can be difficult to *prove* that a material selection was 'suitable' for operation within a particular environment or process fluid.

Figure 7.5 shows several possible chemical corrosion failure mechanisms on our case-study radial fan. One particular form, *crevice corrosion*, results when a local crevice feature prevents the formation of a passivating oxide film (typically on stainless steel). The result is accelerated local pitting on material that would normally be considered corrosion-resistant, especially if small amounts of acidic water (rain or plant run-off) contaminate the bearing lubricating oil.

Galvanic corrosion

This is caused by dissimilar metals in contact in a conducting environment. The mechanism is well described in numerous technical books but, again, there is often the difficulty of *interpreting* published data in the correct way. Published sources normally give accurate values of potential differences between elemental metals but those for real engineering materials are often stated, rather defensively, as 'indicative only'. This means that it can be difficult to diagnose properly potential differences between, for example, types of carbon or stainless steels. Galvanic corrosion is also subject to the unpredictable effects of external influences such as local geometric shapes and variations in electrolyte density, flow, etc. Altogether an uncertain picture. Practically, at the design assessment stage of a failure investigation it is often possible only to categorize possible galvanic problems if adjacent materials are very well spaced in the electrochemical series, such as copper and aluminium. Smaller differences can be difficult to diagnose. Remember, however, that we are working towards defining causation – for which you may not *need* to be able to describe the mechanisms in precise metallurgical or chemical terms.

Figure 7.5 shows a typical galvanic occurrence – in this case it is very localized galvanic corrosion. The aluminium alloy fan blades of the electric motor have been assembled using a soft copper hammer to tap the fan ring into place on its shaft. The copper has a large potential difference from the (almost pure aluminium) alloy fan blade so residual copper has set up a strong galvanic cell in the fan blade assembly. Local galvanic corrosion has produced a pit, producing a crack initiation point for a bending/fatigue fracture. Note the relevance of *timescale* here – the sequence of events was:

- contamination with copper (incorrect assembly method)

then

- the galvanic corrosion process producing the crack initiator

then

- bending fatigue

then

- eventual failure of the blade by a principal stress mechanism.

Note how this has a slightly different emphasis to that of the example of chemical corrosion failure – here the corrosion plays less of a part in the path to failure (you could argue it is only 25 percent of the focus points) but it is still the *initiator*, once the assembly mistake has been made. We will look at this type of argument more fully in Chapter 8.

Electrochemical corrosion
Corrosion by an electrochemical rather than galvanic mechanism can occur whenever sufficient potential exists between contacting or adjacent metals in the presence of a conducting electrolyte. The metals do not have to be dissimilar. It can occur on a macro-scale, in which metal removal is evenly distributed over the surface, or in small localized areas where 'micro-cells' are set up, producing limited areas of potential difference. Such areas have more concentrated metal removal, often producing a shiny, pitted appearance. It is most common in fluid equipment such as pumps, boilers, coolers and hydraulic systems, particularly when water is present in some form. Electrochemical corrosion is frequently found near areas of pure galvanic corrosion – it can be difficult to differentiate between the two mechanisms. In common with other types of corrosion, it acts more as an *initiating* mechanism for other more catastrophic causes of failure rather than itself being a true proximate cause.

Figure 7.5 shows an electrochemical failure mechanism found on rotating equipment; corrosion of bearing components caused by poor earthing. This is more prevalent on high-speed machines and journal bearings which can build up a static charge. This charge can distribute itself unevenly throughout the component geometry resulting in unpredictable corrosion of journals or mating faces. The components become anodic with respect to earth, and so corrode. On highly stressed components this can easily initiate a further failure mechanism.

Stress corrosion cracking (SCC)
SCC is really a hybrid failure category. It occurs when a combination of tensile stress and (normally electrochemical) corrosion act together to cause surface cracks. These, in turn, act as initiators for later crack propagation, mainly by fatigue. This is one type of 'corrosion fatigue' – another hybrid category. SCC is a common failure category in those applications where surface-passivated materials such as austenitic stainless steels are in contact with a corrosive environment such as a solution containing high chloride concentrations. The passive surface layer is brittle and starts to fracture at low stress, producing fine cracks which are then easily attacked by the corrosive environment.

Figure 7.6   A typical micrograph of stress–corrosion cracking (Courtesy: Royal and Sun Alliance Engineering)

SCC is one of the easiest failure categories to identify. It is characterized by multiple bifurcated cracks, often oriented in the plane of the principal stress. The cracks are bifurcated because they propagate down the grain boundaries – this causes the cracks to 'split off' from their side walls instead of propagating from their tips (Fig. 7.6). Strictly, SCC is considered a hybrid failure mechanism because it is the combination of corrosion and (albeit low) stresses that initiates and opens up the cracks. In practice the technical description of the mechanism is frequently simplified so that SCC is represented as a unique 'proximate' cause in itself. As a failure category, corrosion is one of the easiest to diagnose. From the viewpoint of failure investigations carried out for insurance or commercial liability purposes, this diagnosis of some *corrosion mechanism* is often the key point. Long and complex technical definitions will often confuse the issue and eventually cloud the conclusions of the investigation, instead of making them more precise. Precision is fine, but corrosion technology will not always support precise views – it contains too many opposing theories and technical opinions.

Inherent defects
More poor conclusions are probably built around this category than any of the others. This is due partly to misunderstanding of the metallurgical facts about inherent defects but also, in fairness, because

of the difficult nature of these so-called metallurgical 'facts'. Here are four 'statements' that you will encounter during failure investigations:

- 'Materials, and welds, always contain some defects – it is impossible to get rid of them all.'
- 'You can't diagnose a failure as being caused by an inherent defect that is undetectable – because you can't prove it was there.'
- 'If a material has inherent defects which are smaller than some 'critical' crack size, it won't fail.'
- 'Materials just *don't* contain 'inherent' defects – if they do have them then there has been a design or manufacturing fault somewhere.'

Are all of these statements 'true' in that they represent technical fact, or maybe only some of them? A closer look will reveal some clear technical contradictions between pieces of some of the statements – so they can't, by definition, all be correct. Maybe they are all misconceptions – but if they were, then how would you describe the existence and effect of inherent defects in metals? Fortunately, the answer is clear:

None of these statements are technical *fact*. They are simply theories.

This makes things easier; because they are ideas – mental constructs – they can all be true (or untrue) at the same time, without contradiction. It all hinges on which theories of material failure you believe in.

This has key implication for failure investigations where the prime objective is to work towards *conclusions*, rather than to seek to expand the scientific boundaries of metallurgy. The first step is technical understanding. Forget the notion that there is, somewhere, a single and irrefutable technical explanation of the way that materials *start to fail*. Be careful not to confuse this with the process of how the failure mechanism continues (or *propagates*) which is a different thing. Figure 7.7 shows the generally accepted phases of a failure. Elastic behaviour, up to the yield point, is followed by increasing amounts of plastic flow. During this stage, the material is referred to as 'damaged', because of the irreversibility of the plastic deformation. Note how this is different to the common usage of the term 'damage', which infers that a component is in some way broken. The *fracture* of the material starts from the point in time at which a crack initiation occurs and continues during the phase of crack propagation. Finally, the material breaks. The correct terminology for this is actually 'rupture', although this has a few slightly theoretical implications as to the way that the material actually parts – so the term *breaks* is perhaps a better one to use. You will also

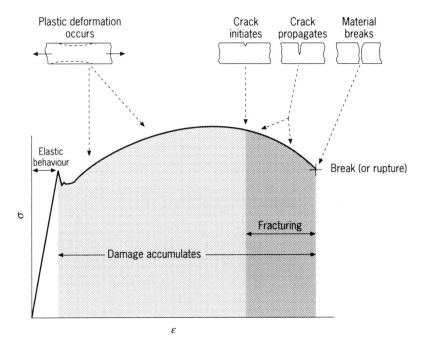

Figure 7.7   The three stages of 'failure'

see the action of breaking referred to as 'fracture' – this is not wrong, although, as shown in Fig. 7.7, fracture is more of a process, rather than a single incident.

The beginning of the failure process – initiation – is the most important stage and also the most difficult to understand. In essence, there are two main theoretical approaches that you will need. I suspect there are many more, but that they are even more complicated and probably have less robust empirical 'backing' than do the two main ones. Remember that they are *models* of how failures start – nothing more. The key difference is the assumption made about whether nominally homogeneous materials always contain pre-existing defects or not:

- The *no defects* ('uncracked') approach. This assumes that a material is entirely homogeneous, and does not contain any defects which are *significant* in an engineering context.
- The *pre-existing defects* ('pre-cracked') approach. This assumes that a material, although still nominally homogeneous, will always have pre-existing defects that are capable of growing. The existence of

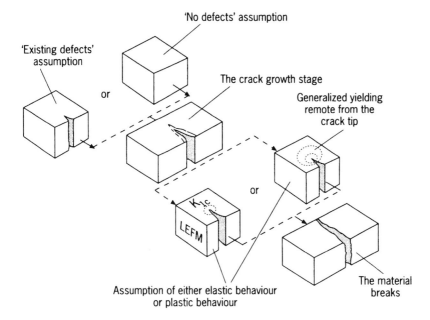

'No defects' assumption

'Existing defects' assumption

or

The crack growth stage

Generalized yielding remote from the crack tip

or

K_{1c}

LEFM

Assumption of either elastic behaviour or plastic behaviour

The material breaks

Points of agreement (in all approaches) are:
- Crack size is important: large cracks cause quicker failure
- Cracks always grow: they never get smaller
- All damage is cumulative

Figure 7.8 How metals fail – different *approaches*

crack initiators is a characteristic of the material – they are already there.

Note also the different approaches to crack propagation (see Fig. 7.8). The differences are to do with assumptions about how the material behaves at the tip of the crack. The two alternative approaches here are:

- The material remote from the crack tip behaves in an absolutely elastic way. This encourages the use of the so-called LEFM (linear elastic fracture mechanics) method of analysis with its related concept of the fracture toughness (K_{1c}) parameter.
- Material remote from the crack tip has yielded, leading to a complex set of happenings referred to generically as 'plastic collapse'.

For the practical purpose of defining the *cause* of failures in engineering materials, the 'pre-cracked' assumption leads to all manner of difficulties. The concept that all metallic materials contain, by definition, pre-existing cracks leads to the natural conclusion that all components are a failure waiting to happen and that as all components have these cracks, then these cracks cannot, sensibly, be the proximate cause of failure for those few components that *do* fail. In theory, the initiation phase would disappear from consideration and failures would be attributed to the subsequent propagation or fracture phases of the damage mechanism. Clearly, the technical rationale for such an argument is damaged by the existence of numerous other components living in the same 'crack propagation conditions' that have *not* failed. This is a weak position. So:

If you subscribe to the 'pre-cracked' approach to material failure, some of your failure investigations will be *inconclusive*.

This is a danger – you will often see it in very detailed investigations where the depth of analysis is sufficient to expose these weaknesses in the unanimity of metallurgical understanding. The 'pre-cracked' assumption might well be a fine and persuasive theory (it may even be *true*) but it won't help you in commercial, accident, or insurance-related failure investigations.

The 'uncracked' concept is more helpful. It encourages the search for a reason for the crack that initiated the process of failure and helps define the purpose and objectives of failure investigations. It also fits in well with the formal definitions of proximate cause and failure 'events' that are meaningful to insurers and commercial parties that commission failure investigations – they can work with the answers. The approach has three interesting technical characteristics that you should use. They are not an essential part of the theory, but on a practical level they make things easier:

- The (awful) term *'crack-like defects'* is used extensively to describe cracks and other types of defect that can act as failure initiators. Defects such as porosity are not cracks, but in this context they do *act like them* – hence the terminology.
- Crack-like defects are a *surface phenomenon*. This has a strong empirical basis; cracks can exist wholly in the body of materials (castings are a good example) but experience shows that fractures nearly always initiate from defects that meet the surface of the material because this is where stress levels tend to be highest.
- Crack initiation is a *macroscopic mechanism*, so defects that initiate cracking are visible under low magnification, or even with the naked eye. Again, this is a practically orientated statement – if defect identification were to be extended to the micro-level, the inevitable existence of microscopic crack-like features such as dislocations takes you away from the 'uncracked' approach and back to the

assumption of pre-existing defects. Opinions vary as to the limits of the macro-level. Practically, nearly all relevant crack-like defects will be easily visible at a magnification of $\times 50$. More than half will be visible at $\times 25$.

What are these defects?

They can take many forms. Figure 7.9 shows the main ones that you will find in practical failure investigations. This is one area where it is difficult to be specific about which of them are the most common – things are heavily dependent on the type of material, the geometry of the component, and the nature of its operational environment. Note the various *sources* of the different types of surface defect shown; some are formed during the manufacture of the material while others rely on help from an external influence such as an imposed corrosion mechanism or stress regime. There are many others that are not described individually in Fig. 7.9, so don't treat this figure as a full inventory. We can look at them briefly; they are well covered in metallurgical textbooks if you need more technical details.

Actual cracks

The location of macroscopic cracks can be caused by numerous factors: stress concentrations, weld 'decay' (in stainless steel) or poor homogeneity of the material. Figure 7.10 shows a typical defect in a weld. Even the so-called 'micro-cracks' caused by stress-corrosion or other complex mechanisms are, strictly, still classed as 'macro', because they are visible at a magnification of less than $\times 75$. Note the comparison with the 'critical' crack size needed to cause propagation. You have to be a little wary of this stated size (it varies from material to material and even between reference sources) – cracks can soon grow and *then* act as initiators.

Mechanical damage

Surprisingly small incidents of mechanical damage can act as crack-initiators. There is no need for metal to have physically been removed from a surface – small amounts of metal transfer in the form of burrs or marring marks form excellent surface stress-raisers that can initiate cracks. This also extends to the more general issue of surface finish. All metal machining and finishing processes (forming, milling, grinding, etc.) produce mechanical damage to a surface. It is well known that rough-finished surfaces, say more than $12.5\,\mu m$ R_a, have significantly less fatigue resistance than ground surfaces of $0.1\,\mu m$ R_a and better. This is because the rough surfaces contain more crack initiators. Hence

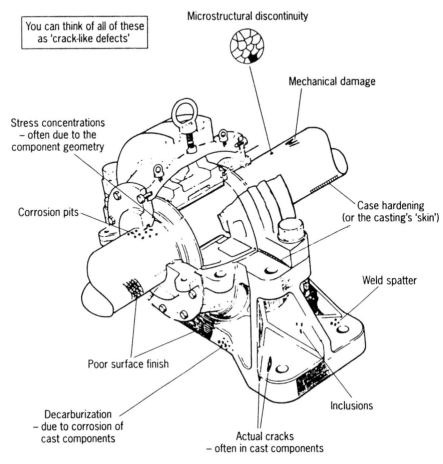

You can think of all of these as 'crack-like defects'

Microstructural discontinuity

Mechanical damage

Stress concentrations – often due to the component geometry

Case hardening (or the casting's 'skin')

Corrosion pits

Weld spatter

Poor surface finish

Inclusions

Decarburization – due to corrosion of cast components

Actual cracks – often in cast components

Figure 7.9   Typical types of 'crack-like defect' that can initiate failures

the actual size of a burr or physical mark needed to act as a crack initiator is very small – so visible 'macro' damage can be a danger. Weld spatter has a similar effect.

Macrostructural discontinuities
Discontinuities occur when the grain structure varies over the surface of an engineering material. The most common type is that where part of a surface has been hardened (flame, case-hardened or nitrided) for design purposes. Hardening changes the metallurgical structure and cracks can occur at the junction of treated areas with the unhardened parent material. Another example is where segregation occurs of carbon present in the matrix. Again, many of the defects will be visible at a macro level, even though their origin may be microstructural.

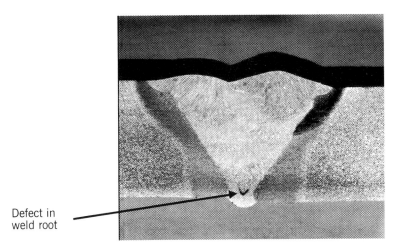

Defect in
weld root

Figure 7.10   A typical defect in a weld (Courtesy: Royal and Sun Alliance Engineering)

Corrosion pits
These can be caused by any of the three main corrosion mechanisms described earlier. The corrosion pits inevitably have some sharp edges and the mechanical strength properties in surrounding areas may be reduced by the ongoing corrosion activity. Cracks initiate easily under such conditions. Decarburization of carbon-rich materials such as cast irons is a related activity with similar detrimental effects.

Inclusions
An inclusion is the existence of a particle of dissimilar material to the parent metal matrix. It can result either from material precipitating unexpectedly out of solid solution or simply by contamination by an impurity in the form of another metallic or non-metallic material. Tungsten (from cutting tools) and copper are often found as inclusions in high-alloy steels. As well as the stress concentration problem resulting from the lack of homogeneity, inclusions can also set up galvanic cells within the parent material, which makes the situation worse.

As a summary, you should now see why it is necessary to be careful when using the term 'inherent defects'. The 'uncracked' approach assumes that materials, by definition, do not contain pre-existing cracks. This infers that any crack-like defect that you do find must have been introduced by one of the mechanisms I have explained. We are now, by looking at what happens at the very beginning of the failure process, getting nearer to being able to define accurately the cause of a failure.

We are starting to address *causation*. This shows the importance of this activity – the search for crack initiators. Finding reasons why they are there is one of the key skills of effective failure investigation.

Fatigue

Fatigue is a common feature of engineering failures, and Figure 7.1 showed how around 80 percent of failures involved a fatigue mechanism of some sort. Fatigue is an *imposed* failure mechanism in that it needs an external source of stress cycling in order to operate. Mechanical stress due to imposed vibrations, and stresses caused by thermal cycling, can cause it. Fatigue can play two roles in the process of material failure. It is the predominant mechanism by which existing cracks propagate through a material section but it can also act as a crack *initiator* in the earlier stage of the process (Fig. 7.7). The ways in which these activities occur are far from straightforward but they are generally accepted – at least at the level of description that we need. This is useful – it helps encourage clear and incisive statements on causation if there is not too much diverse technical opinion to take into consideration.

What is it?

Fatigue is the result of imposing repeated cyclic stresses on a material containing either no cracks or existing crack-like defects (those two theories again). The stresses may be at a level well below the static yield stress. Fatigue can either initiate crack-like defects, or propagate those that are already there, as previously described, and can be caused by almost any type of stress regime (bending, torsion, etc.) as long as it is cyclic, producing fluctuating stresses. It can be either 'high-cycle' or 'low-cycle', depending upon the frequency. Figure 7.11 summarizes the main characteristics of the fatigue mechanism. Further technical explanations are quite standard, and you can find them in almost any book on materials. From our point of view the important characteristics are those that impinge directly on the various parts of failure investigation, particularly the conclusions. It is worth paying particular attention to this area because fatigue is perhaps the most important failure *category* – you will find it appearing again and again in failure investigation reports, and in your own investigations. Sadly, you will also see failure investigations where fatigue becomes a source of confusion – it can easily encourage indecision and poor quality conclusions. This is not because it is not understood, in a technical sense, by the metallurgists and engineers involved, but because it has to be properly explained in a way that has *relevance* to the specific objectives of the investigation. This is not difficult (once you understand the technique), largely because the technical characteristics of fatigue *are* so well documented and accepted. Figure 7.12 outlines the main technical characteristics that are relevant to failure investigations. Look at these points carefully, they form the 'backbone' of many of the technical causation statements that we will discuss in Chapter 8. Try to treat the content of Fig. 7.12 as your main 'learning points' about the mechanism of fatigue failure. Don't waste time digging too deeply into

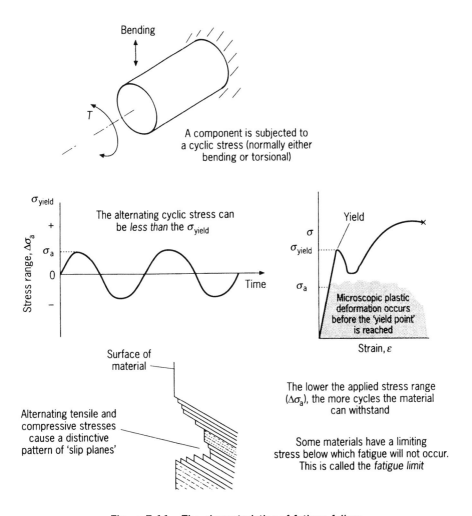

Figure 7.11   The characteristics of fatigue failure

APPEARANCE?

Fatigue fractures normally have a smooth but visibly stepped surface. The crack propagates without causing much plastic deformation until 'arrested', giving the characteristic visible 'beach marks'. Under higher magnification, the surface may show striations, caused by individual stress cycles.

INADEQUATE DESIGN?

Not necessarily. A material does not have to be 'overstressed' to suffer from fatigue. Fatigue can frequently occur at stresses as low as 30 percent of principal yield stress (σ_y). This is due mainly to 'stress raisers' on the material's surface raising the local stress above σ_y.

WHICH MATERIALS?

All of them. Some materials are designed to try to resist fatigue but all metals suffer to some degree. Carbon steel has a clear 'fatigue limit' – a stress level below which it has (in theory) infinite resistance to alternating stress cycles.

STRESS CONCENTRATIONS?

Fatigue mechanisms thrive on changes of section, sharp edges, slots, small radii, and any other feature that produces an area of stress concentration. Millions (literally) of engineering components world-wide have failed because of this.

INITIATOR?

Doesn't need one. A fatigue mechanism is capable of initiating its own 'crack-like defect' and propagating it. Fatigue can therefore exist at several points across the 'timescale' of a failure event.

Figure 7.12   Fatigue failure – the *relevant* points.

the metallurgical aspects of it – it may be interesting, but further analysis won't necessarily help you perform better and more conclusive failure investigations.

Figure 7.13 shows a classic fatigue of a shaft. The initiation point at a surface defect can be clearly seen. The fatigue crack has propagated

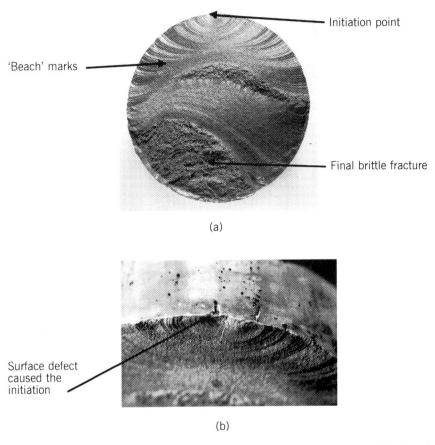

Initiation point

'Beach' marks

Final brittle fracture

(a)

Surface defect
caused the
initiation

(b)

Figure 7.13 A classic fatigue, initiated from a surface defect (Courtesy: Royal and
Sun Alliance Engineering)

across approximately 65 percent of the shaft section with final failure by
brittle fracture.

Brittle fracture

Brittle fracture is, strictly, a type of principal tensile stress failure but it
has such distinct characteristics that it is best considered as a separate
category. It is also quite common, being responsible for many 'classical'
failures of engineering components. Brittle fracture is characterized by
crack propagation without significant plastic deformation or reduction
of area of the material around the fracture face. In contrast to ductile
fractures, the fracture faces tend to be fairly flat. They are generally of

light, speckled appearance, and may show lines or striations across the surface.

Brittle fracture can be encouraged either by physical or metallurgical factors. The main ones are:

- *Microstructure.* Large grain size produces a brittle material, as does excessive hardness owing to incorrect heat treatment.
- *Environment.* Chemical environments can cause embrittlement, particularly in a high tensile strength material.
- *Low temperatures.* Although basic steels are usually ductile at ambient temperatures, all steels become more brittle as the temperature reduces. There is a well-defined transition temperature (it varies between material grades), below which a metal will fail predominantly by brittle fracture.
- *Stress concentrations.* This is more a contributory factor than a direct cause. Although brittle fracture can be propagated by principal stresses, it is often found to have been triggered by a stress concentration. Notches in rotating shafts are particularly good at acting as initiators for brittle fracture.

Timescale

In its purest form, brittle fracture happens almost simultaneously with the initiation of a crack-like defect. Unlike fatigue, it cannot initiate such a defect itself, so it can't act as its own initiator. For practical purposes, however, the actions of initiation and propagation are instantaneous. This has important implications for failure investigations – it is almost implied in the technical description of brittle fracture that the component was operating in a 'satisfactory' manner, immediately before it failed. There is no failure 'timescale', as such. Sub-transition temperature failures of structural steels are perhaps the best example of this. Figure 7.14 shows an example of brittle fracture of a stepped shaft.

The second type of brittle fracture failure is that which occurs at the final rupture of the material, following either a fatigue (as in Fig. 7.13) or a ductile fracture mechanism. Some amount of brittle fracture can be found on fracture faces of even straightforward principal tensile stress specimens. Brittle fracture surfaces are flat, bright, and crystalline in appearance, in contrast to the duller, more contoured surface appearance of a ductile fracture where the material has deformed plastically. This plastic deformation causes work-hardening – one reason why the final rupture happens by brittle fracture. In this case, the failure does have a timescale, with the final brittle fracture being almost consequential to the crack propagation that has gone before. This has very

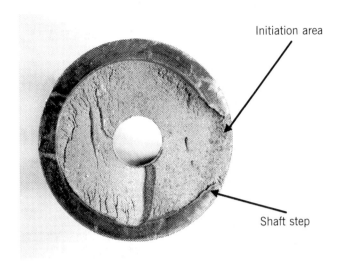

Initiation area

Shaft step

Figure 7.14   Brittle fracture – in this case of a stepped shaft (Courtesy: Royal and Sun Alliance Engineering)

different implications for failure investigators – it would obviously be wrong to conclude that the component was operating satisfactorily immediately before the brittle fracture occurred, because it was already damaged (as in Fig. 7.7) and waiting to fail.

It is easy to confuse these two examples of the brittle fracture mechanism. When you categorize a failure as 'brittle fracture' you can see the importance of describing properly *how* it occurred. This is a good example of the point made in Part I of this book; that an understanding of the *context* of an engineering failure is often just as important as the cold, metallurgical definition of the failure mechanism. Effective failure investigation is about *why*, as well as 'how'.

Creep

This is a temperature-induced failure mechanism. Creep is the long term result of tensile stresses on material held at elevated temperatures. The material deforms plastically to the point where it ruptures. It happens mainly to pressurized components; such as boiler superheater headers and tubes which operate above about 400 °C. At these temperatures, the metal suffers thermal softening via a 'dislocation climb' mechanism and voids start to appear in its grain structure. The voids grow and coalesce until rupture occurs. The problem is particularly bad in thin-walled vessels – these have an inherent tendency to become unstable and can suffer from bulging at high temperatures as the material starts to creep.

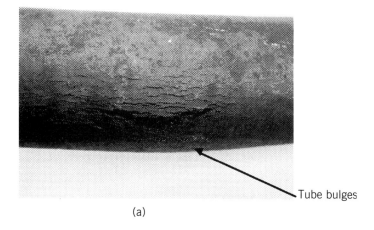

Tube bulges

(a)

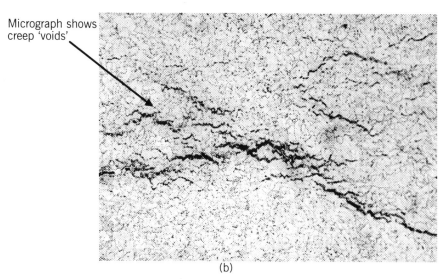

Micrograph shows creep 'voids'

(b)

Figure 7.15   Typical creep damage to a boiler superheater tube (Courtesy: Royal and Sun Alliance Engineering)

The situation is shown in Fig. 7.15. Theoretically, there are two separately identifiable forms of creep process:

- *Intergranular* – in which the creep voids appear at the boundaries of the grains. This happens at relatively low stresses, around $\sigma_y/3$.
- *Transgranular* – this is caused by higher imposed stresses, the voids appear within the grains, giving a more ductile rupture mechanism.

In practice, there is little need to differentiate accurately between these two types. They are both categorized as a creep and have a similar effect on the integrity of the component.

Timescales
You can think of creep as being almost totally time-dependent, with small temperature or stress differences having only a secondary effect on the lifetime of a component. The creep/time relationship of steels is well documented in technical literature – you will see references to empirical 'steady state creep rates' at various temperatures. From a failure investigation viewpoint, there is little disagreement over the conclusion that creep is a gradual and progressive failure process. It happens over time. This means that an engineering component will likely have been operating in a weakened condition for some time before the actual failure (rupture) occurs. There is clear technical evidence for an acceleration in creep strain rate immediately preceding rupture but this is mainly of metallurgical rather than practical interest. Creep failures are sometimes caused by poor material choice (i.e. a *design* fault) or by excessive operating temperatures (an *operations* fault), and sometimes both. The good thing about creep is that it is easily definable, and difficult to confuse with other failure categories.

Impact failure
This is more a sub-category of failure than a category in itself. The metallurgical aspects of impact failure are identical to those of the relevant failure mechanism involved (tensile, pure stress, bending or whatever), but the actual nature and timescale of the failure are different. Impact failures are common in engineering components because impact has the effect of exaggerating 'static' loads by a significant factor. Dynamic loadings on most equipment components which move or rotate in some way can easily produce stresses of six or seven times those experienced under static design conditions. The situation is worse in highly loaded components – these may be designed using high tensile materials in which the ductility is lower than for normal 'utility' materials. This lack of ductility decreases the ability of the material to absorb the shock loadings, so it is more likely to fracture.

From a failure investigation viewpoint, there are two main uncertainties that make impact failures so common: design and materials. Accurate *design* of engineering components, so that they will confidently resist all impact loads, is difficult. The effects of features such as local stress concentrations are notoriously hard to predict – however accurate the design analysis there is *always* some unpredictability in the way that a particular shape will resist impact loads. Components such as gear

teeth, splines, and reciprocating engine components are good examples.
The second source of unpredictability is *material properties*. Material
can be designed to have increased impact-resistant properties, as measured by Charpy or Izod tests, and while this technology is reasonably
well understood, the technical nature of impact resistance is such that it
is less predictable than the other mechanical properties. You can see this
in almost any Charpy test; three separate specimens have to be used to
calculate an average value and it is not unusual to see a 30–40 percent
variation between the individual specimens' measured values. Factors
such as grain structure, heat treatment, and the location of the specimen
in the parent material can cause the impact values to vary. The message
is simple: impact resistance of individual engineering components is
difficult to predict, which is maybe why there are so many impact-
related fractures.

Technically, impact is most likely to result in a failure which has
predominantly brittle fracture characteristics. Ductile ones are rare,
because of the speed of the crack propagation. This reaffirms the view of
impact failure as a sub-category of failure – you can think of impact as
verging on the subject of *why* the failure occurred (next chapter) rather
than being a pure technical description of the failure mechanism,
explaining *how* it happened. The sources of impact loadings are
themselves varied:

- *Internal* sources – these come from the equipment itself, normally
  from overload, high vibration (acceleration or velocity) levels, or
  from dynamic components such as clutches, couplings, and brakes.
  Effects are often consequential – it is not necessarily the component
  that caused the impact load that will break.
- *External* sources – these are more obvious but equally unpredictable
  factors such as wind and seismic loads and abnormal events such as
  collisions.

Timescale
In failure investigation terms, pure impact failure is instantaneous. The
crack is initiated at the moment of impact and propagates in milli-
seconds to rupture, i.e. true *catastrophic* failure. In practice this is not
always what happens – it is more common for the initial impact to act as
the initiator of a crack which then propagates either by principal stresses
or, more likely, by fatigue. In theory at least, impact failures should
never happen, but the reality is different: There are few types of
mechanical equipment where *all* the stressed components can be
designed with a large enough factor of safety to resist expected and

unexpected impact loads. Design compromise is necessary, bringing with it the inevitable risks.

The next step

We have looked in this chapter at the main categories of engineering failure – you are likely to see all of these at some time. Remember that categorization is the technical step of recognizing and defining a failure mechanism, so it is about *what happened*, rather than why. Take another look at Fig. 7.1; this pattern of the frequency of the various categories is a reasonable approximation, perhaps surprisingly so, to what you will find over a large number of investigations. The prevalence of fatigue is almost guaranteed in 70–80 percent of cases. Even failure investigations become easier if you know what to expect.

The next step is the big one: *causation*. I will make the important point again, for the sake of good order:

Effective failure investigation is about *causation*, not just metallurgy.

It is the technical aspects, those of failure categorization, that enable you to capture the correct descriptions of the cause of a failure, on which so much depends.

KEY POINT SUMMARY: FAILURE CATEGORIZATION

1. Failure categorization is the technical step of *recognizing and defining* a failure mechanism – it is not about expressing it.

2. It is not easy, but becomes easier if you know broadly what to expect (look at Fig. 7.1).

3. The main categories of failure are:

- wear
- corrosion
- principal stresses
- fatigue
- creep
- inherent defects
- impact (which is really a sub-category).

4. *Fatigue* features in about 80 percent of failure cases.

5. Materials science (metallurgical) knowledge plays a part in categorizing failure – but it is not the full story. Treat it *selectively*.

Chapter 8

Deciding causation

About causation

The issue of *causation* is what divides good, effective failure investigations from the mediocre or just plain poor ones. Basically, it is about whether or not you can be decisive:

EFFECTIVE FAILURE INVESTIGATIONS REACH ACCURATE AND DECISIVE CONCLUSIONS

while

POOR ONES DO NOT

Poor, indecisive conclusions (or no real conclusions at all) are an all-too-common feature of failure investigations. Perhaps there is a reason for this. Conclusions are difficult, technically complex, and almost guaranteed to be controversial in some way, because they apportion responsibility and liability. This is why indecisive or 'hedged' conclusions can look attractive, but mainly to those who write them. In reality, many of their advantages reside only in illusion. There is little real value, to anyone, in an inconclusive failure investigation unless it is being carried out from a pure research viewpoint (the type D investigation outlined in Chapter 2) – otherwise, all it does is waste somebody's money. If you want to do this – fine: but it's best not to expect your clients to come flocking back. This is more to do with economics than any particular business philosophy – there is intense competition among indecisive failure investigators because they are not in short supply.

In this chapter we will look at this issue of causation, trying to focus on the technical elements that matter, rather than those which are peripheral. We will look at how best to express causation in the form of logical and structured technical statements that other people will be able to understand. The case study of the radial fan, introduced in previous chapters, is used, so you can see the principles used for a real example. First, some key points on decision-making.

Decisions

Decisions are the currency of good failure investigations. Fortunately, these decisions are linked firmly to technical disciplines so they are nothing like as complex or paradoxical as those in perhaps management, or logistics, or medicine. Causation decisions are still not fully quantitative; there are no neat columns of digits to add up to give you that single answer. What there *is*, however, is a robust technical backing which enables decisions to be made. Figure 8.1 shows one way of thinking of it – note how at least 70–80 percent of the decision 'field' is well supported by various aspects of technical *fact*. There are areas where there are choices between viable alternatives (judgement) and one where there are questions of interpretation (the shaded slice), but these are relatively small compared to the area that is backed up by factual observations about the failure. The message of Fig. 8.1 is that if you get hold of all the available facts of a failure, then the majority of the difficult decisions have effectively been decided for you and are recorded in precedent, technical literature, and other reference sources. All you have to do is pick them up. This needs a certain degree of technical competence (and confidence) – knowing where to look or who to ask.

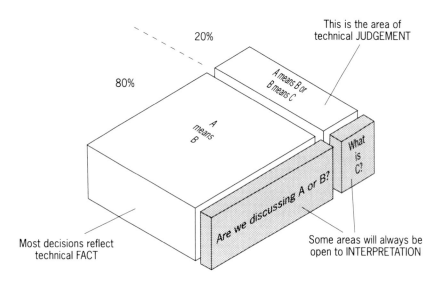

Figure 8.1   Most causation decisions can be reinforced by technical fact

Knowledge and experience will give you this, as will practising making causation decisions and learning from the feedback. To summarize – there is nothing magical about making causation decisions. If you concentrate on building up your technical knowledge base and don't shy away from using it, they will start to follow naturally.

Causation statements – structure

Causation statements consist of simple steps meshed together within a coherent structure. This gives a technical commentary which has the property of logical progression, each statement leading on logically from the one before it. Causation statements are not purely technical – they can contain issues about what happened (events) and when (timescale), and statements about activities, and people, that were at fault. Figure 8.2 shows the basic content. Note that the elements shown are not options; they all have to be there if the overall commentary is to be of full value. Can you also see how this fits in with the overall framework of damage and causation outlined in Part I of this book? (Take a quick glance at Fig. 2.2 if necessary.)

The case study: more details
We will look at each step in conjunction with our case study example of a radial fan that failed. The analysis will follow on from Fig. 6.4, which gave various first impressions of the failure. Figure 8.3 shows details typical of the type of preliminary information you would have following your initial failure investigation visit. Figure 8.4 is the result of your subsequent design and operational appraisal (Chapter 6), and Figs 8.5(a) and 8.5(b) show details of the main shaft failure. Note how the information is starting to build up, and how each of these three figures

---

A CAUSATION STATEMENT NEEDS TO COVER:

- Information *about* the failure.

- Technical details of the failed equipment or component.

- Facts about *what* happened, and *when*.

- Very precise explanation of the *causes* of the failure.

Figure 8.2   Causation statement – content

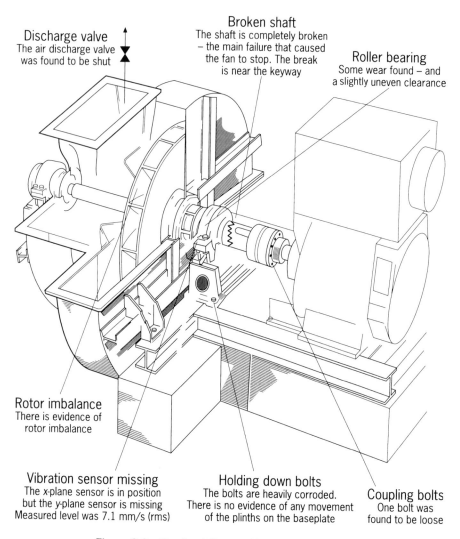

Discharge valve
The air discharge valve
was found to be shut

Broken shaft
The shaft is completely broken
– the main failure that caused
the fan to stop. The break
is near the keyway

Roller bearing
Some wear found – and
a slightly uneven clearance

Rotor imbalance
There is evidence of
rotor imbalance

Vibration sensor missing
The x-plane sensor is in position
but the y-plane sensor is missing
Measured level was 7.1 mm/s (rms)

Holding down bolts
The bolts are heavily corroded.
There is no evidence of any movement
of the plinths on the baseplate

Coupling bolts
One bolt was
found to be loose

Figure 8.3   The fan failure – this is what you found

contributes something to the story of the failure. The so-called 'main' failure is clearly the shaft because it was this breakage that caused the fan to be reported as not working. The failure had serious consequences – the air was needed for a complex chemical process in order to prevent the process chemicals solidifying and they have all gone hard. Also, the valve between the fan and the process was found to be shut when you made your initial inspection visit. During this visit you were plagued by the plant maintenance engineer; who followed your every move:

DESIGN APPRAISAL

Your checks	The results
Impeller balance grade check (from last maintenance)	Compliant with technical standard ISO 1940 G40 – but this is a low balance grade
Torque resistance of the broken shaft (using $T/J = \tau/r = G\theta/l$) principal stress formulae	The shaft has a factor of safety of >3
Bearing sizes and clearances	Satisfactory – based on well-proven 'handbook' data

OPERATIONAL APPRAISAL

Your checks	The results
Overcurrent trip setting	Correctly set – so no evidence of overloading
Maintenance records	Satisfactory – all planned maintenance activities are up to date
Vibration readings	Maximum recorded reading of 7.1 mm/second r.m.s. – slightly higher than the design value of 6.3 mm/second
Process air shut-off valve	Records show how it was closed after the fan failure

Figure 8.4   The radial fan study – Design and operational appraisal results

'I told them [the management] that shut-off valve was no good – all done to reduce cost, you see' ...

He continued:

'I guess the spindle must have broken, the valve slammed shut and the

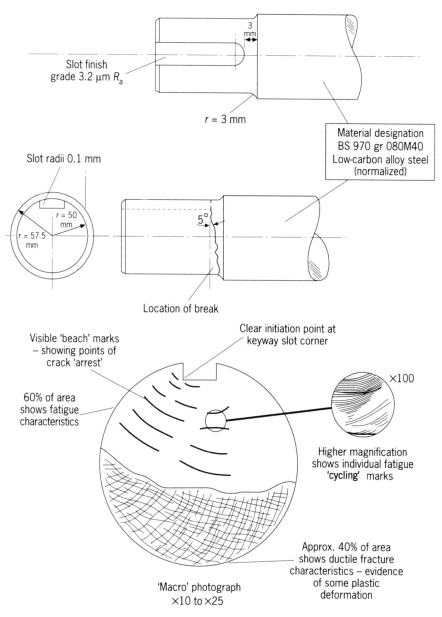

Material designation
BS 970 gr 080M40
Low-carbon alloy steel
(normalized)

Slot finish
grade 3.2 μm R_a

$r = 3$ mm

Slot radii 0.1 mm

$r = 50$ mm
$r = 57.5$ mm

5°

Location of break

Visible 'beach' marks
– showing points of
crack 'arrest'

Clear initiation point at
keyway slot corner

×100

60% of area
shows fatigue
characteristics

Higher magnification
shows individual fatigue
'cycling' marks

Approx. 40% of area
shows ductile fracture
characteristics – evidence
of some plastic
deformation

'Macro' photograph
×10 to ×25

Figure 8.5(a)   The fan shaft failure – technical details

resulting back pressure overloaded the fan, snapping the shaft. The fan itself isn't very good quality either, my guys are always telling me it sounds like a bag of nails. Pity I didn't check it myself, if you see what I mean' ... [which you don't].

Summary of design features

Shaft surface finish: 1.6 μm R_a Shaft manufacture: Forged and turned
Keyway surface finish: 3.2 μm R_a Keyway cutting: Slot-milled
Shaft step radius: 3.0 mm
Keyway corner radii: 0.1 mm

Mechanical properties	Condition	Microstructure
R_e = 300 MN/m^2 R_m = 570 MN/m^2 Hardness = 180 HB Impact = 20 J A = 20%	Normalized	Pearlite and ferrite

Properties compliant with BS 970 grade 080M40 (normalized)

The broad technical diagnosis is:

Figure 8.5(b)   The fan shaft failure – technical details

One of the plant operators, however, tells a different story:

'Corroded bolts my friend, corroded bolts – see those holding down bolts?' ... [you do] ... 'that's pure chemical corrosion attack, caused by alkaline vapour from the chemical process. That's a maintenance responsibility though, and they don't like to do it because it takes nearly two hours just to get the coupling guard off so you can access the bolts – and it's pretty hot and dirty down there. It's obvious what happened, the bolts came loose, the alignment shifted and the shaft broke'.

The final slant was placed on the failure by the fan manufacturer's representative, who had to be almost prised off the telephone to speak to you. He seemed frightened by the mere mention of almost *any* technical conclusion, just in case it pointed to the failure somehow being 'his fault': He wasn't particularly technically forthcoming either:

'I suspect faulty operation because, you see, the evidence seems to be that it was overloaded' ... [you haven't actually seen this evidence but he obviously has] ... 'and any machine will fail if it is operated outside its limits. This is a precision-designed product which complies with all international design standards and was made under an ISO

9001 quality assurance system. Anyway we've made hundreds of these and the others are fine'.

The structure

You can see how the situation is becoming a little confusing. Although the broad technical diagnosis of the failure is fairly straightforward (Fig. 8.5), there are a number of technical observations (in Fig. 8.3) and verbal reports that are complicating the issue. If all you produced as your 'causation diagnosis' was a commentary on the details in Figs 8.3 to 8.5, then the situation would *remain* confusing. Further steps are required.

Figure 8.6 shows the structure of a causation statement – note how it follows a logical progression, as described earlier. We will look at the five elements in turn relating each to the fan case study.

Step 1: Preliminary facts
This first element of the causation statement should contain mainly technical facts about the equipment. It can be quite short. These are background technical facts which would be equally valid if there had not been a failure. The content is as follows, broadly in the order shown:

- *A short introduction*: this explains what the equipment is and what it does (its 'function'). Mention model type, serial numbers, and the location and date of manufacture.
- *Technical 'design' information*: this needs to be more detailed than that available from manufacturers' catalogues. It should mention:

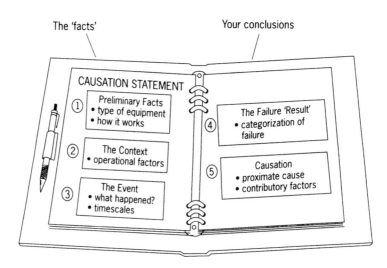

Figure 8.6   The structure of a causation statement

– technical standards
– design lifetime of critical components.
Some of this information will be available from the equipment data sheet but most will come from the results of your detailed design assessment (note the various information sources shown in Fig. 6.3).

• *Fitness for purpose issues*: it is important to include information which shows how (or whether) the piece of equipment is suitable for the function that it has to perform. Mention any particular features such as sealed bearings, special balancing requirements, design margins, etc. that set it apart from other models from the same manufacturer. Such information is often shown in the equipment's original purchase order. Again, it can be identified during a thorough design assessment.

• *Materials of construction*: materials are worth a separate mention, particularly those used to make the components that have failed. You may have to do a little research, starting with the general arrangement drawing and then using published materials hand-books. Accurate identification as to which materials were used is essential, if the later technical stages of failure categorization and causation are to have proper relevance.

These technical facts are preamble but they also have a high level of technical *precision*. This is a common weak area – many failure investigation reports only provide broad, rather obvious, information lifted from the front page of the equipment sales catalogue. A good factual technical preamble should be tailored to the specific features of the equipment in question. This will make the later steps easier and more *robust*. Figure 8.7 shows two contrasting examples related to the fan case study.

Step 2: The context
'The context' explains how the equipment was operated before the failure. It is still part of the factual description of the situation, but concentrates on the equipment's *use*, rather than its mechanical design. Don't confuse this with general technical preamble – operational effects are at least a contributory factor to many engineering failures. Try to treat the context as of equal importance as the mechanical design. This part of the causation statement often contains important implications for the reason for failure. It is best to stick to facts, rather than express opinions (these come later).

Part of your 'context' comments should mention whether the equip-ment was being used in the way *for which it was designed*. This can be a

Example (a): Weak

The failed item is a single stage radial fan No XXX/123 stated to be manufactured by the Imperial Fan Company. No specific technical standards are shown on the nameplate. The original purchase order asks for a design life of 150 000 hours and the requirement that 'all relevant international standards shall be complied with'. The fan is of standard design for process air delivery at 3 bar. The shaft is of carbon/manganese steel with all other items in low or medium carbon steel, painted as required.

Example (b): Better

The failure occurred on a single stage single speed (1500 r/min) radial fan supplying air at 3 bar (gauge) to a chemical process. The fan operates for 24 hour/day. There is no parallel-connected or redundant fan in the system. The fan is type RF1 serial No XXX/123 (data sheet ref. DD/454 rated at 250 kW (electrical). The manufacturer is the Imperial Fan Company – they supplied the fan/coupling/base assembly only – the motor and air valve were supplied separately.

Design review shows that the fan is designed to be compliant with BS 848. This covers mainly testing and does not specify materials or factors of safety for rotating components. Nominal design life is 150 000 hours, however the bearings are the only components with 'rated' design lifetime. The shaft is not 'fatigue-life' rated. The shaft is keyed. The 'design' impeller balance grade is ISO 1940 G6.3. Impeller clearance and general manufacturing tolerances are shown on GA dry FF.678 Rev. 1.

The fan has no 'special' design features required for air duty. Design criteria such as balancing, alignment, and factors of safety are the same as for other fans in the manufacturers' range.

Materials of construction are:

Keyed shaft:	Material BS 970 type 080 M40. This is a low carbon alloy steel used in the normalized condition. It has a nominal yield strength (R_e) of 280 MN/m^2 and a hardness of about 150–200 HB.
Other components: (casing, base, bolts, etc.)	LCS grades. There are no particular strength or impact property requirements for these components.

Note how the shaded parts add *precision* to the technical description.

Figure 8.7   The radial fan failure – Step 1: Preliminary facts

complex issue, but an important one – you can expect to hear claims that equipment was operated outside its design limits in most failure investigations. This is one of the reasons for a thorough design and operational assessment as part of the failure investigation process (Chapter 6). Four specific areas need to be addressed in this part of the causation statement:

- *The operating regime*: i.e. whether the equipment operates for 24 hours/day, or only intermittently. Total operating hours are also important. If running hours are not recorded, you will have to make an estimate.
- *Loadings*: these are always a relevant factor, but particularly if the equipment has been overloaded or its loading is very cyclic, or unpredictable. You should mention if the equipment is likely to be subjected to *imposed* loadings, as a result of process conditions somewhere else in the plant.
- *Monitoring*: have the various operating parameters of the equipment been monitored and recorded? Motor current speed, vibration levels, and bearing temperatures are the common ones for rotating equipment.
- *Maintenance*: have maintenance activities been carried out on the equipment in compliance with manufacturers' recommendations (and good engineering practice)? Mention maintenance records, if they exist.

These four areas form the skeleton of this part of the causation statement. Don't make this part too long – certainly not more than 200 words – but try to make your statements *pointed*. Figure 8.8(a) shows a good example, and Fig. 8.8(b) a rather poor one, containing some points to avoid.

Step 3: The failure 'event'
This is the third part of the causation statement shown in Fig. 8.6. The objective is to describe accurately the event, or events, that occurred when the failure actually happened. It has to be a description based on fact rather than opinion. It also needs to have the correct *focus* – that of the engineering investigator (you) – rather than merely reflecting the story according to one or more of the involved parties. You can see earlier in this chapter exactly how different and confusing some of these can be. *Look* at it from your own perspective – *you* describe the events. The concept of *focus* was discussed in Part I of this book – turn back and read it now if you find it a little elusive. Watch also the *type of event*

Example (a): Good

The fan is operated at rated speed for 24 hours/ day on 'full load' air delivery. Stop/starts are rare. A recorded total of 12 020 running hours accrued before failure, over a calendar period of 19 months since installation. It is estimated that there were less than 100 stop/starts in this time. The external atmosphere is corrosive owing to the fan's location in the process plant. This operating regime is within the 'design philosophy' of the fan.

Operating regime

The fan cannot be overloaded as it is of centrifugal (dynamic displacement) design. Process fluid cannot feed back through the air ducts as the head is limited by fixed-length sparge pipes in the process vessel. Similarly, it is not possible for any fluid 'shock loads' to reach the fan. Mechanical shock loads are limited by the use of a star-delta 'soft start' arrangement on the motor. The impeller/shaft cannot be physically locked. Overcurrent protection devices and vibration trips are installed to prevent the fan operating outside its design limits.

Loadings

Motor current and speed are continuously monitored on a meter. Bearings have one thermocouple each. Although designed to incorporate 2-plane (x and y) vibration monitoring, this fan has x-plane (axial) monitoring only. Parameters are logged manually every hour.

Monitoring

There is a planned maintenance system for the fan bearings, coupling and mounting bolts. Basic records are kept, showing only when items are replaced.

Maintenance

Figure 8.8(a)   The radial fan failure – Step 2: Context

Example (b): Weak – with some points to avoid (shaded)

Points to avoid

The fan normally operates at full load except during maintenance periods. The local meter showed 12 020 hours running at the time of failure. This is not an onerous operating regime for a radial fan.

This comment is opinion rather than technical *fact*.

The fan is prevented from overload by an overcurrent trip. This makes it unlikely that overload was the cause of the shaft failure.

Don't *prejudge* in this part of the causation statement.

Sparge pipes are used to deliver the fan output to the process vessel. It is not known whether these will allow pressure surges, which cause fan overload.

Totally unhelpful comment (what happened to your design assessment?)

Operating parameters are measured by various sensors installed on the fan. Vibration is measured in the *x*-plane only.

A review showed that the maintenance programme is incomplete. Individual bearing wear and clearances are not recorded.

Possibly an *unjustified implication*. There is no real evidence (yet) that the maintenance activity is incomplete. This comment could be misleading.

Figure 8.8(b)   The radial fan failure – Step 2: Context (*cont.*)

that you describe; concentrate on tangible physical events that would be observable during the failure, rather than obscure metallurgical happenings. Metallurgical explanations are better at explaining events rather than describing them. For example:

- 'The shaft fractured' is a possible, accurate attempt at *description*.
- 'The crack propagated instantaneously' is a metallurgical, as well as physical, observation. It is an attempt at *explanation*, rather than description.

If in doubt, the test is always to ask yourself if it would have been possible to see the event happening. If it would, mention the event here.

Timescale is also an important consideration. Several aspects of the results of a failure investigation hang on the chronological sequence of events. In insurance-related investigations in particular, the proportion of the total cost of damage that is insurable can be affected by the sequence in which various bits of the damage occurred. For the causation statement to be useful, divide this step into three chronological periods:

- The period before, and *immediately prior to*, the main failure event. Say what the equipment was doing during this period – was it on load, at full speed, etc? Mention any unusual operating conditions, and whether it was being adjusted by an operator.
- The *event itself*. Make clear whether there was any external influence on the equipment such as operator intervention at the precise moment when the failure occurred. Say exactly what happened – did the equipment stop, overspeed, catch fire, or what? Keep your focus on to this very short time-period; seconds rather than minutes.
- The *sequence of events* that followed the main failure event. Now you can spread the time-horizon out, relating the succession of events that happened next. Your starting point is immediately after the failure event. Some of the sequence will ultimately end up being classed as a part of the main event but other areas will be more obviously consequential. It is difficult to predict just how long the time-horizon will be – consequential effects of, for instance, impact damage may not appear for many operating hours after the impact event.

This third step of the causation statement is important in the way that it starts to control terminology. The terms used need to be consistent with those that will be used in step 4, which will start to close in on failure *categorization*. Chapter 7 showed how very precise, tailored terminology is needed to describe accurately the complex technical aspects of failure mechanisms. This means that you have to use the same terms (if they are needed) when describing the events as you do when you start to categorize – i.e. move from fact to opinion. This is critical because any inconsistency will cause your subsequent technical arguments to lose their strength. Figure 8.9 shows a commentary on the failure 'event' for our radial fan case study example. I have highlighted some 'categorization' terminology, so you can see how it is used.

Step 4: The failure 'result'
The next step is to provide an accurate and detailed description of the failure mechanism(s). This meshes together the findings from the inspection visit (Chapter 5) and your technical conclusions on the failure categorization (Chapter 7). We looked at failure categories separately in Chapter 7: however, it is more than likely that this section will involve describing several of them together. They rarely appear in 'pure' form, except in very simple failures.

This step is also about *selection*. In nearly all failure investigation cases there will be some engineering observations and findings made by yourself, and others, that are only peripheral to the failure mechanisms. You have to select those that are directly related, and that will help support the overall technical causation statement. You will see this in the case study example. A few other points are important:

- *Terminology*. Use the accepted 'failure category' terms, as mentioned earlier. You are aiming for a close technical description of the mechanisms of failure because loose descriptions at this stage can jeopardize the effectiveness of the entire investigation. Remember that the robustness of some of these technical descriptions will inevitably be questioned by other parties.
- *Consistency*. Your descriptions of the results of the failure obviously need to fit in with the various parts of the failure event(s) you have identified in step 3 of the causation statement. Don't introduce any contradictions, or new events you haven't mentioned before.
- *Logical progression*. We have seen this requirement before. Make sure that each technical statement follows on logically from the previous one. You are aiming for close, dense technical dialogue – but keep it simple.

The results of some failure investigations can become quite complex – they are not all a neat succession of simple events. Part of the purpose of the three points above is to keep everything in order, when the situation

The failure occurred at 12:31 a.m. on Tuesday 02 July. The fan had been operating continuously at 100 percent load since the last start, approximately 138 hours before the failure. During this period, motor current readings were logged at their normal values. Vibration (x-plane) readings were constant at 7.1 mm/sec, which is higher than the 6.3 mm/sec design level. The last planned maintenance activities were completed four months previously, as scheduled – no excessive wear or corrosion had been reported at this time.

Before the event

No process or operational changes were being made at the time of failure – everything was constant at full process load. No shock loadings were imposed on the fan and there was no operator intervention.

Immediately prior to the event

The first indication of failure occurred at 12:31 a.m. when the 'loss of process air' alarm No. 4A/167 sounded. The fan tripped approximately 5 seconds later.

The event itself

The subsequent sequence of events was:

- The vibration alarm sounded during the rundown.
- Fan stopped smoothly approx. 10 seconds after trip. No bearing seizure.
- An operator went immediately to check the fan (12:24 a.m.). There was no fire. It was apparent that the fan shaft had failed.
- The process was shut down and the air valve closed manually at 12:39 a.m.
- The visual check and damage report was made at 05:00 a.m. after it got light.

The following sequence

Note: Specific failure categorization terms are shown shaded.

Figure 8.9   The radial fan failure – Step 3: The failure event

does get complex. By following the principles of consistency, using correct terminology and, in particular, by ensuring that your text has logical progression you can produce a good clear causation statement. It will be robust. People will understand it. It will also fulfil the requirement for demonstrating *causality* (showing how the result followed from an event). This is the absolute core of failure investigation: engineering discipline. Figure 8.10 shows good (and bad) examples of failure 'result' commentary for the radial fan case study.

Step 5: Causation
Causation is about opinion and fact meshed together so that they explain what happened. The description must also make good technical sense. The main purpose of this step is to home in on the *proximate cause* of the failure. It is the identification of proximate cause that drives the solution to a failure investigation, enabling it to be settled. It is difficult to over-emphasize the importance of this final step – your conclusions on proximate cause will have implications for allocating responsibility and liability for the failure. They will identify what (and probably who) was at fault. There are only three simple guidelines for a good commentary on proximate cause:

- *Clarity*. We have discussed previously the need for a failure investigation to reach decisive conclusions. This becomes much easier if the causation statement is *clear* as well as technically correct.
- *Based on evidence*. The commentary has to be linked to the objective technical findings of the investigation. Some too-specialized failure investigators can formulate immaculate causation statements that have little relation to the actual failure. They use the same causation statements for different investigations because they are technically consistent and sound good.
- Based on the *category of investigation*. Part I of this book outlined the four main types of failure investigation (we called them A, B, C, and D for convenience). The causation statement needs to fit the correct context of the investigation. It is of limited use phrasing your comments for a product liability lawyer if the investigation has been commissioned by the equipment users. They need to know different things to fit their *focus*.

These are three exacting requirements – so the final parts of a good causation statement are far from being easy. Even the most practised failure investigators can provide commentaries which, with the best will and intention, are incomplete or lack direction. We will look in some

(a) GOOD

The fan shaft broke radially, in the region between the end of the rectangular key slot (motor end) and the change of section of the shaft. The crack initiated at the area of stress concentration at the corner of the key slot. A stress concentration is likely in this area owing to the sharp edge of the slot and the proximity (only 3 mm away) of the small (3 mm) shaft section radius.

The crack propagated via a high-cycle fatigue mechanism (bending and/or torsion) for approximately 60 percent of the shaft diameter. The final rupture was the result of a brittle fracture mechanism, caused by excessive principal (torsional) shear stress on the remaining shaft section. This final rupture was what was observed during the failure 'event' recorded at 12:31 a.m. on 02 July.

The fractured shaft showed no evidence of mechanical wear or corrosion or external impact damage. Macro-examination showed no evidence of internal defects such as inclusions. The micrograph showed a uniform grain structure across the shaft section. Surface finish in the key slot was measured at 3.2 μm R_a using a comparator. Spectrographic examination of the material showed it complied with specification.

(b) BAD

The fan shaft fractured because of a high-cycle fatigue mechanism (*not strictly true*) in the region of the key slot. This occurred catastrophically at 12:31 a.m. on 02 July (*no, it didn't*). The material was found to be compliant with specification (*which one?*) and the micrograph was acceptable (*but what did it show?*). Fatigue needs an initiation point before it can start – this appeared to be the sharp edge of the key slot (*partially true but a flawed explanation*).

Note: background technical findings are shown in Fig. 8.5.

Figure 8.10   The radial fan failure – Step 4: The failure *result*

detail at the radial fan case study, showing how the final step of the causation is developed, and some of the decisions behind it.

Radial fan case study: Step 5
Figure 8.11 summarizes the situation of the failed radial fan. The left hand columns show the information and technical conclusion reached already in earlier steps – note how the findings related to the main failure (the broken shaft) are shown shaded. The other findings (the closed air valve and corroded mounting bolts) are clearly not directly connected to the failure. Note also the findings identified as 'contributory factors' – these are related to the failure but are not to be confused with the definition of the *cause*. The three-stage failure result shown is a common finding in failure investigations on rotating machinery, or any other equipment where there are imposed vibration or fatigue conditions.

The most important information is that in the 'causation' column at the right hand side of Fig. 8.11. Note how the failure has been broken down into the identifiable stages that occurred *before* the failure event. The purpose is to show how the failure did not start from the event when the fan shaft finally broke – this was, in fact, the *final* stage of the failure. Again, this is a common feature of machinery failure. It is often also misunderstood. If you mistakenly treat the final shaft failure as the *only* timescale of failure, it is almost impossible to decide what caused it because the cause was active some time before. You have to look backwards through time in order to make technically accurate statements about causation. Using this technique, the allocation of cause (the far right hand section of the figure) becomes easier. You now have the key to that most elusive of findings; which is:

PROXIMATE CAUSE.

Proximate cause was introduced in Part I of this book. Originally an insurance-related term, it is equally useful in other categories of failure investigations because of the way that it picks out the happening, or situation, that is nearest to being the fundamental reason *why* the failure occurred. Proximate cause is well defined in insurance case law as the cause which is nearest the failure in *effect*, rather than in timescale. You can see this in action in Fig. 8.11 – look how the fatigue mechanism which propagates the crack is nearer the breakage event in time, but not in *effect*. If there had not been a stress concentration (caused by the design) to start with, then the fatigue alone would not have resulted in this particular failure. The relatively small deviations from impeller

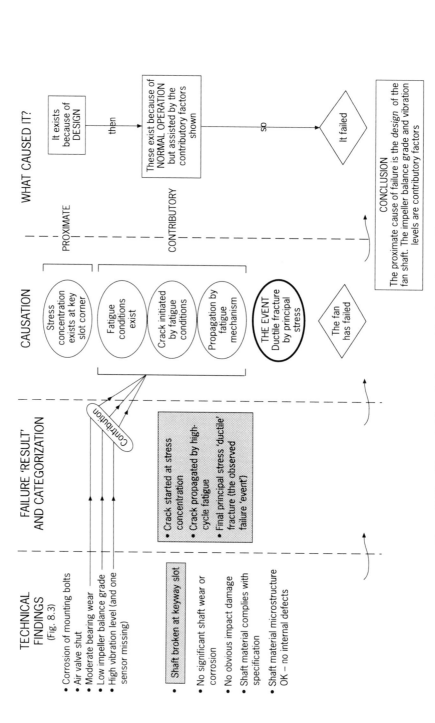

Figure 8.11   The radial fan study – working towards *causation*

balance grade and maximum vibration limits tell us that the imposed fatigue condition was mild, and therefore acted as a contributory factor only. It did not break the shaft by an impact load of the type discussed in the last chapter. The contributory factors are also separately identified in Fig. 8.11, to complete the picture.

Once causation has been decided, then it has to be written in commentary form, as the last part (Step 5) of the causation statement. Figure 8.12 is the result.

Some of the points in Fig. 8.12 are most applicable to an insurance-related failure investigation – you will see them used in Chapter 10 of this book which looks at specific aspects and further specialized terminology used in insurance damage claims and settlements.

These previous five steps, added together, make up the causation

---

The main event that caused the failure was the breakage of the drive shaft, during normal operation of the fan. The *proximate cause* of failure was inadequate design of the shaft in the region of the rectangular drive key slot. The combination of sharp corners and rough surface finish of the key slot – and its proximity to the small-radius change of shaft section – produced high stress concentrations. These were sufficient to result in crack initiation under fatigue conditions. The crack propagated to failure under the fatigue conditions experienced during normal operations in rotating machines. This was a single and unbroken chain of events.

The following were contributory factors to the failure but were not, either singly or together, sufficient to be considered as the proximate cause:

1. Low impeller balance grade (ISO 1940 G40). This is a design responsibility.
2. The high operating vibration (7.1 mm/sec). This would be caused predominately by impeller imbalance.
3. Bearing wear. This is part of 'wear and tear' expected during normal operation of the fan.

Other technical observations had no significant effect. There is no evidence that any other external influences or operator actions were involved in the failure.

Figure 8.12   The radial fan failure – Step 5: *Causation*

statement for the failure investigation. This is your *product*. Figure 8.13 shows the assembled statement, comprising approximately 1000 words. Aim for about this length: any longer and it could lose its technical clarity.

STEP 1: PRELIMINARY FACTS

The failure occurred on a single stage single speed (1500 r/min) radial fan supplying air at 3 bar (gauge) to a chemical process. The fan operates for 24 hour/day. There is no parallel-connected or redundant fan in the system. The fan is type RF1 serial No. XXX/123 (data sheet ref. DD/454 rated at 250 kW (electrical). The manufacturer is the Imperial Fan Company – they supplied the fan/coupling/base assembly only – the motor and air valve were supplied separately.

Design review shows that the fan is designed to be compliant with BS 848. This covers mainly testing and does not specify materials or factors of safety for rotating components. Nominal design life is 150 000 hours, however the bearings are the only components with 'rated' design lifetime. The shaft is not 'fatigue-life' rated. The shaft is keyed. The 'design' impeller balance grade is ISO 1940 G6.3. Impeller clearance and general manufacturing tolerances are shown on GA dry FF.678 Rev. 1.

The fan has no 'special' design features required for air duty. Design criteria such as balancing, alignment, and factors of safety are the same as for other fans in the manufacturer's range.

Materials of construction are:

Keyed shaft:	Material BS 970 type 080 M40. This is a low carbon alloy steel used in the normalized condition. It has a nominal yield strength (R_e) of 280 MN/m^2 and a hardness of about 150–200 HB.
Other components: (casing, base, bolts, etc.)	LCS grades. There are no particular strength or impact property requirements for these components.

Note how the shaded parts add *precision* to the technical description.

Figure 8.13   Radial fan failure – the complete causation statement

STEP 2: THE CONTEXT

The fan is operated at rated speed for 24 hours/day on 'full load' air delivery. Stop/ starts are rare. A recorded total of 12 020 running hours accrued before failure, over a calendar period of 19 months since installation. It is estimated that there were less than 100 stop/starts in this time. The external atmosphere is corrosive owing to the fan's location in the process plant. This operating regime is within the 'design philosophy' of the fan.	Operating regime
The fan cannot be overloaded as it is of centrifugal (dynamic displacement) design. Process fluid cannot feed back through the air ducts as the head is limited by fixed-length sparge pipes in the process vessel. Similarly, it is not possible for any fluid 'shock loads' to reach the fan. Mechanical shock loads are limited by the use of a star-delta 'soft start' arrangement on the motor. The impeller/shaft cannot be physically locked. Overcurrent protection devices and vibration trips are installed to prevent the fan operating outside its design limits.	Loadings
Motor current and speed are continuously monitored on a meter. Bearings have one thermocouple each. Although designed to incorporate 2-plane (x and y) vibration monitoring, this fan has x-plane (axial) monitoring only. Parameters are logged manually every hour.	Monitoring
There is a planned maintenance system for the fan bearings, coupling and mounting bolts. Basic records are kept, showing only when items are replaced.	Maintenance

Figure 8.13   *(cont.)*

STEP 3: THE FAILURE 'EVENT'

The failure occurred at 12:31 a.m. on Tuesday 02 July. The fan had been operating continuously at 100 percent load since the last start, approximately 138 hours before the failure. During this period, motor current readings were logged at their normal values. Vibration (*x*-plane) readings were constant at 7.1 mm/ sec, which is higher than the 6.3 mm/sec design level. The last planned maintenance activities were completed four months previously, as scheduled – no excessive wear or corrosion had been reported at this time.

Before the event

No process or operational changes were being made at the time of failure – everything was constant at full process load. No shock loadings were imposed on the fan and there was no operator intervention.

Immediately prior to the event

The first indication of failure occurred at 12:31 a.m. when the 'loss of process air' alarm No. 4A/167 sounded. The fan tripped approximately 5 seconds later.

The event itself

The subsequent sequence of events was:

• The vibration alarm sounded during the rundown.
• Fan stopped smoothly approx. 10 seconds after trip. No bearing seizure.
• An operator went immediately to check the fan (12:24 a.m.). There was no fire. It was apparent that the fan shaft had failed.
• The process was shut down and the air valve closed manually at 12:39 a.m.
• The visual check and damage report was made at 05:00 a.m. after it got light.

The following sequence

Note: Specific failure categorization terms are shown shaded.

Figure 8.13 (*cont.*)

STEP 4: THE FAILURE 'RESULT'

The fan shaft broke radially, in the region between the end of the rectangular key slot (motor end) and the change of section of the shaft. The crack initiated at the area of stress concentration at the corner of the key slot. A stress concentration is likely in this area owing to the sharp edge of the slot and the proximity (only 3 mm away) of the small (3 mm) shaft section radius.

The crack propagated via a high-cycle fatigue mechanism (bending and/or torsion) for approximately 60 percent of the shaft diameter. The final rupture was the result of a brittle fracture mechanism, caused by excessive principal (torsional) shear stress on the remaining shaft section. This final rupture was what was observed during the failure 'event' recorded at 12:31 a.m. on 02 July.

The fractured shaft showed no evidence of mechanical wear or corrosion or external impact damage. Macro-examination showed no evidence of internal defects such as inclusions. The micrograph showed a uniform grain structure across the shaft section. Surface finish in the key slot was measured at 3.2 μm R_a using a comparator. Spectrographic examination of the material showed it complied with specification.

Figure 8.13   (cont.)

STEP 5: CAUSATION

The main event that caused the failure was the breakage of the drive shaft, during normal operation of the fan. The *proximate cause* of failure was inadequate design of the shaft in the region of the rectangular drive key slot. The combination of sharp corners and rough surface finish of the key slot – and its proximity to the small-radius change of shaft section – produced high stress concentrations. These were sufficient to result in crack initiation under fatigue conditions. The crack propagated to failure under the fatigue conditions experienced during normal operations in rotating machines. This was a single and unbroken chain of events.

The following were contributory factors to the failure but were not, either singly or together, sufficient to be considered as the proximate cause:

1. Low impeller balance grade (ISO 1940 G40). This is a design responsibility.
2. The high operating vibration (7.1 mm/sec). This would be caused predominately by impeller imbalance.
3. Bearing wear. This is part of 'wear and tear' expected during normal operation of the fan.

Other technical observations had no significant effect. There is no evidence that any other external influences or operator actions were involved in the failure.

Figure 8.13   (*cont.*)

KEY POINT SUMMARY: DECIDING CAUSATION

1. Decisions, decisions

A failure investigation that does not reach clear decisions on causation is virtually worthless. The best way to make causation decisions is by focusing on purely technical matters – don't get diverted into other facets of the investigation (interesting though they may be).

2. Your causation statement

A good causation statement forms a key part of your reporting responsibilities. It is a carefully constructed sequence of five logical steps:

1. *The preliminary facts*, giving relevant technical details of the equipment.
2. *The context*: how was the equipment being operated before the failure (and was it what it was designed for?)
3. *The failure 'event'*: this is a *description* – rather than opinion – about what actually happened.
4. *The failure 'result'*: this is also termed 'categorization' – it involves fitting the result into known and accepted technical categories.
5. *Causation*: this part is about what *caused* the failure. Identification of the proximate cause is an important part of the conclusion – particularly in insurance investigations.

Chapter 9

Presenting your conclusions

In the world of commercial failure investigations you can't expect your clients, or other involved parties, to understand fully the nature and depth of your technical investigation into a failure, nor will they be particularly interested in the way that you have done it. They *are* interested in your conclusions. Conclusions are what they are paying for.

It is not uncommon for the *presentation* of conclusions to be the weak link in failure investigations. Conclusions on causation may be clear to the investigator, and in the minds of clients' technical staff, but often remain rather hazy to those with the management or commercial authority to act on the results. Key points about causation, and its implications for liability, can easily be only partly understood – or even worse, misinterpreted. How can you prevent this? It is wise to start with the principle of making your conclusion as simple and direct as you can manage. There is nothing new in this idea – it holds good for many other comparable situations where the objective is to communicate technical information. The other basic rule is always to try and see the *commercial* aspect of your conclusions. They will often hold implications for the financial settlement aspects of failure investigation, so you can rest assured that they will come under close scrutiny. Simply anticipating the general financial aspects is not always good enough, you may have to think in some depth about what the financial consequences for various parties may be. Thinking about money will help your focus and encourage technical conclusions which are tighter and more compartmentalized (see Part I of this book).

On a practical level, technical presentation of the conclusions of an engineering failure investigation separate neatly into three stages:

- the report (the formal written submission)
- the verbal presentation
- the dispute.

Not many books talk about the third one – the dispute – however, it is an undeniable fact that at least half of all engineering failure investigations end up in a dispute of some sort. A technical dispute probably; almost certainly a courteous dispute – but a dispute, just the same (see Part I of this book). As a failure investigator you are absolutely at the centre of any dispute that occurs, so it makes sense to plan for it as part of your presentation strategy.

In most failure investigations these three stages of the presentation will follow in straight chronological order. They are all parts of the same technical story – not different stories (even complementary ones) – but they have different levels of technical *resolution* or depth at which they attempt to explain what happened during the failure. Figure 9.1 shows some other distinguishing characteristics of the three stages. Look how the dispute stage overshadows the previous two in terms of its technical depth, and how the verbal presentation is so shallow. The representation of timescale is also important – the dispute invariably starts during the verbal presentation (before this, people have been kept busy reading your written report) but only starts to attain its true technical depth some time later. These timescales can be surprisingly long – the situation will often still be developing even six to nine months after your first verbal presentation. Many cases don't reveal the true complexity of their technical dispute for a year or so. So remember:

Plan **for the technical dispute – you will rarely be disappointed.**

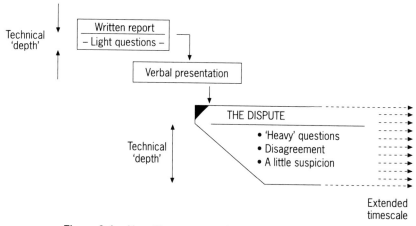

Figure 9.1   Unveiling your conclusions – the three stages

The written report

Your written report forms the 'backbone' of your findings about the failure. It is also the first mechanism for releasing your opinions and conclusion to your client and to the other parties involved in the investigation. It needs to be good.

Written reports can be dangerous – we are all tempted to avoid making conclusions that may later prove to be incorrect. It is natural to want to deal at length with specific technical areas with which we feel most comfortable, even if they are not essential to the conclusion of the report. Sadly, you will see many reports like this. The best way to produce a failure investigation report is to base it firmly on the structure and content of the causation statement, as covered in detail in Chapter 8. Remember the key points of a causation statement:

- A causation statement is not only technical – it also explains what happened (events) and when (timescale).
- It has been built up using fairly tight technical 'rules'; robust assessments of the engineering design and its operating regime.
- The information in it has been sifted for *relevance* (remember Fig. 4.5?)
- It develops in a *logical progression* in five steps:
  1. preliminary facts
  2. the context
  3. the failure 'event'
  4. the failure 'result' (and categorization)
  5. causation.

There is, actually, little more needed to make up a technically complete failure investigation report. The few further additions that are required are there for 'rounding' purposes – to add to the completeness and presentation of the failure story rather than to add any more vital technical substance. They also have an 'enabling' purpose – to emphasize and even repeat some points to help with interpretation. Don't be misled into thinking that the purpose of the additions is to make the report more 'user-friendly', or compliant with some long-lost and dust-laden 'rules of reporting' or similar – these additions are about *effectiveness*. Figure 9.2 shows the general format, and how the extra parts fit around the causation statement. We can look at each of the extra parts starting (for convenience only) at the end of the report.

List of 'evidence' items

This is mainly to help with the level of *formality* of the report. Include at

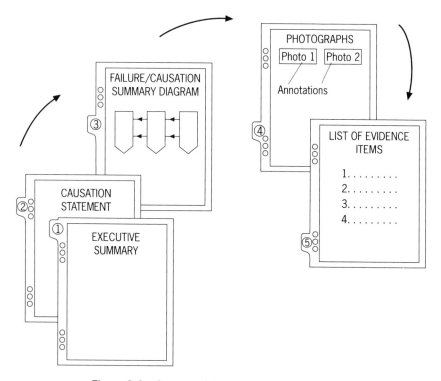

Figure 9.2   Content of the completed failure report

the end of the report a simple list of the items of technical evidence that you found, and used, to help you draw your conclusions about the case. The important items are those that acted as clear technical pointers towards decisions on the categorization of the failure mechanism, its timescale, and the cause. Try to approach each list entry from the viewpoint that your interpretation of it will be questioned later, at the dispute stage. This means you should be careful which items you list – don't include such items as research papers or technical sources that you may have used to obtain general background information, or to get 'a feel' for the subject. You will get your chance to use these later. The list will vary between investigations but the most common technical evidence items you should find yourself using are (in no particular order):

- Photographs of the failed component(s) and other parts of the plant or equipment.
- Technical reports (design reports, inspection reports, etc.) written by other parties, either before or after the failure event.

- Laboratory test reports – mainly about materials analysis.
- Your own results from your design and operational assessment activities (calculations, plant records, and logs).

For an 'average' investigation report expect to have about eight to ten evidence items on the list. If you have four or less, then either it is an unusually simple case, or you have been just too relaxed about the whole affair – take another look. The design assessment activity is the one which is most frequently *under*-investigated.

The failure/causation summary diagram

You need this to help other people comprehend *your* technical understanding of the failure. In short, you want them to see it in the same way that you do – and only in that way. Note that:

THE FAILURE/CAUSATION SUMMARY IS A DIAGRAM, NOT TEXT.

The technique involves representing the failure event, mechanisms (categories), and causation as a series of boxes, joined by links, setting the resulting network against a scale of *time*. You can think of this as a simplified version of the type of 'systems thinking' used in computer programming or risk analysis – the principles are the same. The idea of 'thinking in boxes' is actually one of the background skills of successful failure investigation. The procedural aspects of categorizing the failure mechanism, and getting a clear understanding of the time-dependency of the various stages of failure, are all helped by a bit of low-key 'systems thinking'. In the context of the failure report the purpose of the graphical representation is to explain the time-sequence of events. We will see in Chapter 10 how this is particularly relevant in insurance-related failure investigations.

Figure 9.3 shows a typical model format for a failure/causation diagram. Look how the failure event(s) and mechanisms (results) are shown separately from the causes. This helps prevent confusion – it is all too easy for events, mechanisms, and causation to weld together into a hazy blur. Expressing them graphically helps support the textual explanation in the paragraphs of the causation statement. It brings it all into *focus*. We will see this used later for the radial fan case study.

Photographs

Photographs are a feature of good failure investigation reports. Their purpose is to be informative rather than aesthetic – they are useful in showing the technical details of even the most unphotogenic engineering

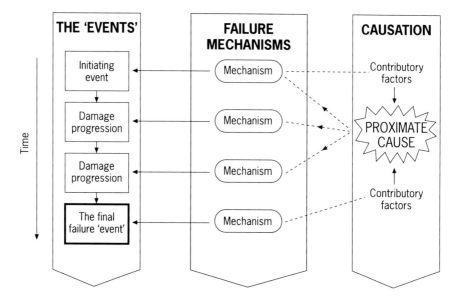

- Remember that this diagram is for presentation purposes only – it is not the place to introduce new technical information
- Figure 9.6 shows a specific example for the radial fan case study

Figure 9.3   The failure/causation diagram – a general model

equipment. Some general rules of using photographs were introduced in Chapter 4, but note the following additional points:

- *Relevance*. It is important that photographs show relevant technical details that help explain or clarify the technical commentaries in the report – particularly the various pieces of the causation statement. Use the minimum number of photographs possible to show the necessary details by choosing frames carefully.
- *Annotations*. Photographs without annotations can be more misleading than informative because readers can either misinterpret the technical message of the photograph or be unsure exactly which part of the photograph they are supposed to be looking at. Annotations should clearly point to specific features of each photograph, forming a link with the technical observations (and of course conclusions) in the causation statement text.
- *Microstructure photographs*. These *must* be explained by detailed annotations or some kind of key. Microstructures are not easily interpreted by non-metallurgists. Material grain size, phases, or

lattice defects can all look much the same unless they are properly highlighted. Use numbered arrows pointing to the particular micro-structural features that you have said are important. Indicate the magnification on the face of the photograph.

- *Location sketches*. Magnified photographs are notoriously difficult to relate to their position on the piece of equipment. Macro- and microstructures vary a lot with location; whether they are from a sample taken longitudinally or transversely in a shaft for instance. In terms of formal evidence, also, photograph locations need to be clear. The easiest way is either to use a location sketch, or to mark-up a large-scale 'general arrangement' photograph, showing where the others in the set were taken. You can use hard-stamp letters, or even chalk marks, on the equipment if this helps make things clear.

Photographs will form an important part of the later visual presentation stage of the investigation, so it is vital that you organize them so that they reinforce your technical conclusions. Nothing is more embarrassing than to have one of your photographs contradict your own finely crafted technical arguments.

The executive summary

All failure reports need an executive summary. Write it *last*, after all the other parts of the report are finished. This is not a book on basic report-writing (you should know how to do this already) but it is worth emphasizing the content of the executive summary, because of its key role.

The purpose of the executive summary is to discourage people from extracting various points and statements from the body of the report text and then misinterpreting them. Readers of the report will need to collect and summarize information for their own reports and discussions and accidental misinterpretation *will* occur, if you don't provide a good enough summary.

Figure 9.4 shows a well-structured executive summary. The most important parts are abstracts from the five steps of the causation statement. The summary is assembled from discrete sentences extracted from each step, with the odd link, if necessary, to help the grammar. A word of warning. When writing your executive summary:

DON'T *REPHRASE* ANYTHING.

Rephrasing can easily cause subtle changes of meaning – these will weaken your technical position – so use exactly the same words as in the report. Don't worry if this makes the summary feel a little staccato – it is

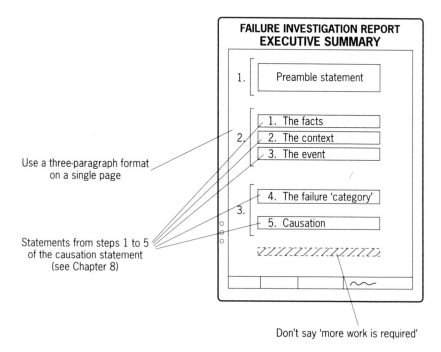

Figure 9.4   How to structure your executive summary

the *content* that must be absolutely right. Figure 9.4 shows the addition of one or two preamble statements. Use these to set the scene, if necessary, but don't waste time repeating what your readers know already. Keep the length of your summary to a single page. You can see an example later, when we look at the completed report for the radial fan case study. Before completing the analysis of report content there are two further, more general, points to consider: terminology and style.

Terminology

It was discussed earlier how much of the strength of a failure investigation report comes from the accuracy and precision of its technical descriptions, particularly those in the causation statement. To achieve this precision it is important to use technical terminology which is both correct and has consistent meaning throughout the report. Terms such as 'failure', 'damage', and 'rupture' are often used too loosely and treated as almost interchangeable (which they are not – they have specific meanings as we saw in Chapters 7 and 8). Categories of failure mechanism such as principal stress, impact, and even fatigue can also sound remarkably interchangeable if you use too much loose, generic, terminology. Sloppy technical terminology is one of the most common weaknesses of failure investigation reports. You will come across it frequently, and can see how it serves to weaken the credibility of a report's causation statements. The message is simple: *be precise* by using the correct technical terms to say what you mean and then *be consistent* in the usage of each term. Pay particular attention to the terms you use in the paragraphs of the causation statement and executive summary – accidental inconsistencies here can reduce the credibility (and commercial value) of your report to a flat zero.

Style

We have all seen too many reports containing some, or all, of the following:

- Marathon 40–50 word sentences that leave the reader gasping for breath but no wiser.
- Dusty Edwardian office-speak, such as:
  - 'therein'
  - 'hereunder'
  - 'herewith'
  - 'it is the author's considered opinion'.
- Quasi-legal mumbo-jumbo like 'the investigated parties do, within fourteen days, wish to exercise their right to ascertain the nature of design review carried out ...'
- The muted clichés of the faded and the indecisive, classic ones being:
  - 'Proximate cause is difficult to determine in this case ...'

or

  - 'In the author's opinion ... although there are of course the academics' views to take into account ...'

Happily, such usages are becoming less commonplace, but you will still see them. They share the common property of looking and sounding *awful*. In fairness, they likely had their place in previous generations of failure reporting – which is firmly where they belong.

Within the constraints of accuracy and precision, try to make the technical style of your reports as light as possible. Stop short of being too colloquial, or flippant, but try to make it easy to read. Avoid extensive metallurgical references or named literature sources in this report, but still try and keep a little academic 'crispness' about it. Finally, a golden rule: it is not good practice to try and copy a particular writing style from other technical or non-technical sources – make the style your own. This is the best way.

The radial fan case study

Figures 9.5 to 9.7 show the executive summary, failure/causation diagram, and list of evidence items. These add to the causation statement shown earlier to form the complete written failure investigation report (excluding photographs) for the radial fan case study.

The verbal presentation

The initial verbal presentation is actually the easy part of a failure investigation. It is normally more shallow (see Fig. 9.1) than the written report and rarely turns out to be the confrontational experience that you may think. In essence, it is your chance to *present* your written report and answer some of the resulting technical questions. At this stage, most of the questions will be to do with clarification, rather than detailed explanations. So, the verbal presentation:

- Provides you with a chance to gauge your client's response to your findings and conclusions.
- Is your opportunity to introduce the concept of *proximate cause*.
- Is *not* the time to ask for relevant technical data (you should have already had this for some time).
- Certainly isn't the time to start having technical arguments about the cause of failure. This comes later.

In most cases your initial verbal presentation will be to your client(s) only, but in others there will be representatives of other parties (perhaps even the 'opposition') present. For this reason it is wise to maintain the verbal presentation as a simple distillation of the facts of your written report. Keep it *predictable* – with no surprises.

The presentation narrative itself should follow the principles of brevity, clarity, and all those characteristics you can read about in books specializing in how to give presentations. Figure 9.8 summarizes the major 'content' points. Note how they are all related to the role of the verbal presentation as a *vehicle* – to relate the content of the written report – rather than as a forum for actually 'solving the failure'.

Questions

Expect questions. Most listeners present at your verbal presentation will like to ask a question or two – if only to show that they have understood

---

THE RADIAL FAN FAILURE: EXECUTIVE SUMMARY

The investigation was performed to find the mechanism and cause of failure of a single-speed (1500 r/min) radial fan. The fan was used to supply air to a chemical process vessel. The fan failed on (date) and the process had to be shut down, incurring significant financial losses.

Para 1: Preamble

The fan (serial No XXX/123) had operated 24 hours/day for 19 months before failure, accumulating approximately 100 start/stops in this period. The nominal design life is 150 000 hours. The fan runs continuously at 100% load and it cannot be overloaded owing to the arrangement of the process piping and valves. The failure event occurred when the fan drive shaft broke during normal operation of the fan.

Para 2: Steps 1–3 of causation statement

The fan drive shaft broke radially, in the region of the end of the drive key slot and change of section (step) of the shaft. This is an area of high stress concentration. The *proximate cause* of failure was the design of this area of the shaft. The geometry enabled a crack to initiate at the key slot corner and propagate to failure. Normal operation of the fan and the low (by design) balance grade of the impeller were contributory factors. Excessive wear, corrosion, or external influences did not contribute significantly to the failure.

Para 3: steps 4 and 5 of causation statement

Note how this text follows the pro-forma structure of Fig. 9.4

Figure 9.5   The fan case study – executive summary

what you said. You can expect most questions to be easily answered at a fairly low level of technical resolution, i.e. without resort to metallurgical reference sources or textbooks. You will often, for some reason, be asked questions on compliance with technical standards, even if they are not *that* relevant to the failure. Answer them to the best of your ability,

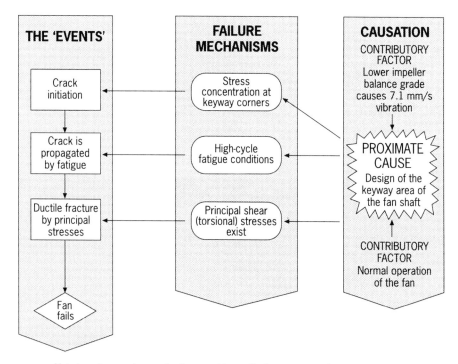

Note how the events, mechanisms and 'causation' are expressed separately but the linkages between them are clearly shown

Figure 9.6   The radial fan case study – failure/causation diagram

but don't introduce anything that is not in your written report. Keep the verbal presentation as a forum for clarification, not new discoveries. This will help protect your position for the forthcoming technical dispute.

The dispute

Disputes occur because of the seriousness of the consequences of your statements on causation – saying what, or who, was at fault. The main issue is normally the cost of the failure, and its subsequent reinstatement. Technical disputes also arise due to *optimism* and its close relative – wishful thinking. The parties at fault in a failure case tend to be generally optimistic that the technical facts behind the causation statements are open to all kinds of interpretation – that they are not technical facts at all. They think that by disputing the conclusions, they

TECHNICAL EVIDENCE ITEMS

1. Manufacturer's data sheet ref: Fan XXX/123.

2. Plant operating logs (dated before the time of failure and on the day of the failure 'event').

3. Relevant design standards BS 848 (fans) and ISO 1940-1, BS 4500, ISO 10816-1 and BS 4235.

4. Purchase order specification ref: Fan XXX/123.

5. Fan shaft material specification ref: BS 970 and reference micrographs.

6. Design/operation assessment results (dated).

7. Inspection visit report (dated).

8. Witness statements (dated).

9. Operators' incident report (dated).

Figure 9.7   The fan case study – 'technical evidence' items

may be able to change the allocation of blame for the failure. Does it work? In most investigations it doesn't. There are very good technical reasons for this – the reality is that the amount of interpretation that can be placed on a *good* set of causation statements is actually quite small. The key to this coherence is the concept of *proximate cause*. Unlike some of the looser ways of expressing 'the cause' of a failure, proximate cause is capable of clear and unambiguous definition. It is also well defined in legal case-law, having been tested over thousands of engineering insurance claims and other failure cases. This gives the following chain of reasoning:

- Causation statements which use, properly, the principle of proximate cause and derive this using robust failure categorization are very difficult to argue against.

because

- There is precious little room for technical interpretation.

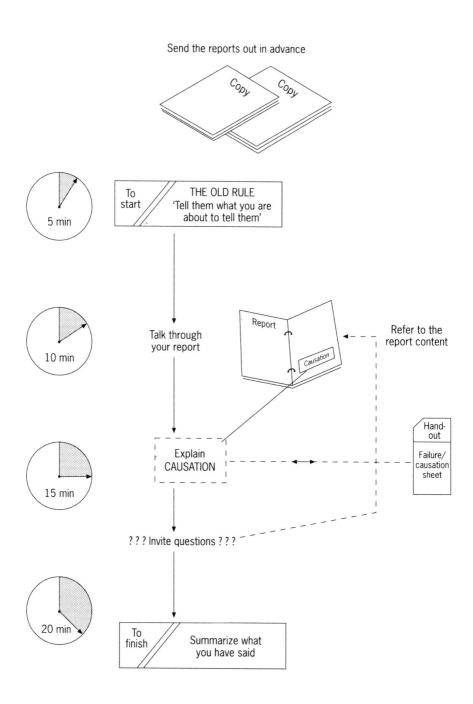

Figure 9.8 The verbal presentation routine

There will always be attempts at interpretation, of course (without them there would be no disputes), but they should not cause you too much trouble, if you have kept to the guiding principles of failure categorization (Chapter 7), causation (Chapter 8) and the pivotal idea of proximate cause.

What does the dispute look like?

Expect it to be quite low key, tinged with a little bit of technical mistrust, rather than the other way round. The main features are generally:

- Extra reporting in the form of statements of clarification, rather than a re-write of your initial written report.
- Answering technical questions (from people that don't understand the idea of proximate cause).
- Discussions and meetings.
- More discussions and even more meetings.

The problem with failure investigation disputes tends to be their timescale – they can go on for much longer than is in any way beneficial (remember Fig. 9.1?).

What happens at the end of it?

Failure investigation disputes end in *settlement* because they have to. As explained in Part I, settlement has a predominantly financial base – it does not have to infer seamless technical agreement on all the statements in your failure report and its additions – and rarely does. What these technical statements do achieve, however, is to *facilitate* the settlement – to *force* the parties to settle. Put bluntly, one side feels the full weight of the technical evidence against it and agrees to end the dispute. This avoids costly legal action or lengthy arbitration. Historically, less than 5 percent of engineering failure disputes have to be settled by legal proceedings in the courts, which makes good sense. You can play your part as a (non-legal) failure investigator in this, by remembering that you are *aiming* for a settlement, and making sure that your technical statements do not make it more difficult than absolutely necessary.

The technical challenge

The best way to think of the dispute stage is as a type of technical challenge. Technical challenge is part of the *scenery* of engineering failure investigation – without it, an investigation would become merely a re-run of well-worn engineering ideas and theories. Challenges involve the spectre of victory and defeat and the concepts of offence and defence (covered in Part I of this book) – things that can sometimes move engineers away from their comfort zone. Seeing the dispute as a technical challenge, almost a type of puzzle with a set routine, can help. Try a positive outlook. We can now look further at some of the skills and 'mechanics' of the dispute.

Basic skill blocks

The causation dispute requires a few basic skill blocks. They are not so different to those discussed in Chapter 4, merely a little more developed to deal with the higher level of technical resolution. There are three, which we will consider in turn.

What you can and cannot present (the code)

There is a code of conduct (absolutely unwritten) that governs what you can present during a failure investigation dispute. The rationale behind this is more practical than ethical – it works to prevent technical errors, misconceptions, and double meanings in both written and verbal submissions. The code is not *law* (you will not find it written down, as I have said) but it does *work*. This means that using it will help your effectiveness, as well as help maintain your credibility in failure investigation. It is summarized in Fig. 9.9. Note how the two sides contain both 'hard' and 'softer' guidelines on what can and cannot be included.

Technical precedent

Technical precedent forms an important part of the 'sub-structure' of defending your technical conclusions. It is most relevant to the *categorization* of the failure mechanism, rather than the resulting decision on causation. Failure mechanisms have been well precedented in thousands of failure cases, so there is little that is new in this area. Precedent, therefore, is *useful*. It is important not to confuse good technical precedent with the many loose comparisons and analogies that are common features of engineering discussions. To be effective, the precedent must be robust – it must be related as closely as possible to the same failure mechanism or identical type of equipment. In practice, this often proves difficult – the variety of equipment types and application is so wide – so nearly all technical precedent tends to be imperfect in some way. The main failure *mechanisms* (categorization), however, are more universal (you often see them referred to as systematic faults, because of the way that they can be accurately identified). Failure mechanisms also

---

YOU CAN PRESENT:

- your honest technical findings (as you see them)
- technical certainties
- facts (but not opinions) gained from your engineering experience
- precedent
- comparisons (similar to precedent but not quite so clear cut).

YOU CANNOT PRESENT:

- incorrect facts (untruths)
- facts which are unsubstantiated
- technical contradictions and apparent paradoxes (there are plenty of both to be found in engineering disciplines)
- prejudices, such as 'fans are always failing', even if that is what your experience tells you
- wild inferences.

Figure 9.9   The dispute – what you can (and cannot) present

have a tendency to repeat themselves. Low temperature brittle fracture of fabricated steel structures such as bridges and ships is one of the best examples – it continues to happen despite having been well understood and documented for at least fifty years. Despite the apparent *strength* of a good technical precedent when used to support your conclusions in a dispute, it is rarely effective as the main element of the technical argument. It can lack power in this role. It is better practice to construct the main thrust of your case from robust engineering knowledge and principles, arranged logically, and let technical precedent play a supporting part.

Dealing with questions

Questions make up quite a large part of the technical meetings and discussions that are an inevitable feature of a failure investigation dispute. Most of the questions will be about your causation statement – any of the five steps in Fig. 8.6 can generate large numbers of questions. The good thing about being subjected to technical questioning is the way that it enables you to retain a level of technical control of the discussion. It is easier to plan for possible questions than to be faced with an aggressive and well-supported allegation that all your technical conclusions are totally incorrect. Strangely (and thankfully), you won't get many of these – they are rarer than you might think – but you will get lots of questions.

There is a simple guideline for dealing with technical questions during a failure investigation dispute:

KEEP LISTENING.
KEEP ANSWERING.

Sometimes you will find the questions run in cycles and you will find yourself being asked the same question you answered last month, or three months ago. Accept this as the way that things work – which is what it is. The depth of your answers is also a key point. It is not wise to argue in greater depth than that used in the question – all this does is open up new technical avenues, increasing the complexity and timescale of the dispute. Keep the answers within the scope of your causation statements *and* within your own sphere of engineering knowledge.

The dispute 'routine'

There are no hard-and-fast rules describing the way that failure investigation disputes will develop – it would be nice if there were. Fortunately, they tend to follow a general pattern, with the bulk of the technical arguments being about one of the four main technical subjects. There is no real reason why one subject has to take priority (there could conceivably be lively technical argument about all four) but it just doesn't seem to happen like this in most failure investigations. The clarity of definition of proximate cause is one reason; it is effective at concentrating the debate (dispute) into a single technical subject related to the failure. Figure 9.10 shows the four subject areas. We can look at them individually and you will see one of them in action in the following further analysis of the radial fan case study.

Disputes about timescale
Timescale becomes an issue mainly in insurance-related failure investigations. The validity of an insurance claim for material (equipment) damage is often affected by whether the failure is deemed to be instantaneous and unexpected ('catastrophic') or whether it can be attributed to a gradually operating cause such as 'wear and tear'. Timescale disputes can usually be solved without too much trouble by accurate reference to the category of failure which follows from the action of the proximate cause. Each failure category (see Chapter 7) has

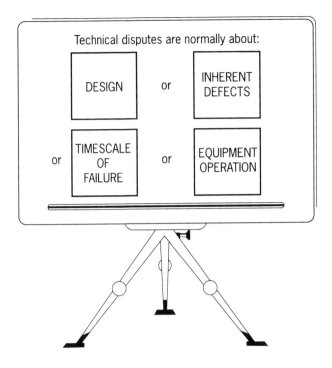

Figure 9.10   The four common subjects of technical dispute

an accepted timescale over which it occurs. This is one of the areas of engineering failure investigation where technical definitions are constructive to the settlement of disputes – for this reason, most disputes about timescale should be short. A short (one page) addendum to the written report is often all that is needed to explain more fully (if you have not done so already) the timescales of the particular categories of failure that have been identified.

Disputes about 'inherent defects'

This can be an insurance-related or straightforward product liability issue. It is common for failure investigators to be commissioned on the loose claims of one or more parties that the failed equipment was 'defective' because it had an 'inherent defect' of some sort (usually undefined at this stage). This demonstrates the effect that terminology can have on the conclusions of failure investigations. In strict technical terms an 'inherent defect' is an identifiable pre-existing crack, inclusion, or similar that can act as a crack initiator – which is a much tighter definition than the common-usage way that you will hear 'inherent defects' discussed in the early stages of an investigation. We discussed in Chapter 7 the different theoretical approaches to material failure – the pre-cracked versus not-pre-cracked assumptions that govern the way material failure is understood and described. You can see why the question 'did the failed component have an inherent defect?' can be difficult to answer in a way that is easily understood by non-engineers. Here is one useful guideline:

It is easier to demonstrate that there *was* a manufacturing defect than to prove there was not.

In practice, because of the different possible metallurgical assumptions, it is just about impossible to prove that a component did not have an 'inherent defect'. This has important implications for your technical dispute – it is more efficient to spend time discussing any defects which *have* been found than getting involved in long theoretical discussions about those that can't be found but which 'may be there'. The whole situation will deteriorate very rapidly if you let it. Put your effort into finding inherent defects (using micro/macro analysis), rather than discussing the search for a single 'law' of material failure. There isn't one (so you won't find it).

Disputes about equipment operation

It is surprisingly *un*usual for equipment operation alone to be the proximate cause of failure. Mechanical equipment, by design, contains safety features (safety valves, shear pins, mechanical interlocks, over-speed protection, etc.) included specifically to guard against the worst cases of operator error. It is much more common for failures to be the result of an omission in the operation regime, such as lack of preventative maintenance or exceeding the design life of replaceable components (excessive wear and tear). Conversely, operation is frequently a contributory factor to a failure, even though it may not warrant, in itself, the classification of proximate cause.

Most disputes in which operation *is* defined as proximate cause centre around the issue of operational *transients*. The common ones are prolonged overloading, shock (impact) loads, overheating, excessive speed, and 'out of design' temperatures. Although the engineering effects of these transients are normally obvious, it can be hard to link the effect directly to a particular operational 'event'. The causal link is very weak, which makes technical *proof* quite difficult to find – an outside observer could be justified in concluding that the evidence is purely circumstantial. Firm documentary evidence from log sheets can help the situation but it is not unknown for equipment to fail because of operator intervention (shutting a valve, etc.) without the action being 'recordable' on even a fully computerized data-logging system. Gearbox failures are a good example of this – bearing and gear teeth failures can occur without any prior warning from the data-logging system or alarms.

The net result is that disputes about equipment operation have a tendency to become a little *unreal*, causing claims about what may or may not have happened to become increasingly hypothetical. Failure investigations are rarely 'solved' like this. The best advice I can give is to make a practice of only discussing equipment operation if there is sufficient documentary evidence to make the line of enquiry worthwhile. Cases based predominantly on colloquial evidence (people's unspecified recollections of what they saw or heard) often founder, and have to be retrieved by concentrating on a different aspect. Sometimes this will mean having to revise your previous firm conclusions on proximate cause, which can become messy. Frankly, conclusions leading to detailed discussions about equipment operation are best avoided, if at all possible.

Disputes about design

This is perhaps the most common form for the dispute to take. Whenever an engineering component fails, someone is likely to ask the question 'was it properly designed?' Disputes about design are invariably longer, more technical, and more expensive than other types. Their only redeeming property is that they are usually *solvable* – as long as you understand the characteristics of the design process itself. There are three of these:

- *Design is mutidisciplinary*: this means that different specialists' skills are needed. So there are going to be several different viewpoints about even simple design features.
- *Design is iterative*: forget the notion that there is always a series of logical incremental steps from A to Z, ending in 'the design'. This means that design steps are not necessarily *traceable* by assessment (so it is difficult to find faults).
- Although mechanical design is based on codes and standards these do not cover everything. The detailed design of some common pieces of engineering equipment is actually influenced very little by published technical standards.

All of these three points indicate imperfections in the process of mechanical design. This conclusion – that design is *imperfect* – provides the key to solving design-based technical disputes. In fact the settlement *only* becomes possible if you accept that the design process is imperfect. The finding is reinforced by looking at the pattern that design-based disputes normally take. I suspect that this pattern is not absolute, or perfect, but I have found it surprisingly consistent over a large number of engineering failure investigations. Figure 9.11 shows the pattern. Note how it has only two simple chronological steps: the design assessment stage and the comparison stage. There are three main implications of this pattern (it is based on observation, remember, not theory):

- Make a thorough job of the *design assessment* before the dispute stage (see Chapter 6). Most (80 percent) of the conclusions are found here.
- Move as quickly as possible to the *comparison* stage. This will save everyone's time and money.
- Plan to settle the dispute by using the comparison activity. Your design assessment is the facilitator of the settlement but its power lies mainly in the process of eliminating possible design errors rather than finding them.

Maybe this sounds a little conceptual, but it seems to represent what actually happens at the end of disputes. It will also happen if you *don't* plan for it, but it will take much, much longer and the cost will rise dramatically. This is not the most effective way to do it. With this in

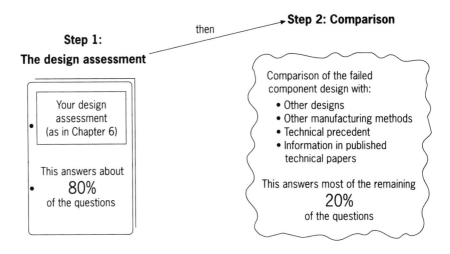

Figure 9.11   The general 'pattern' of design-based disputes

mind, we can now return to the radial fan case study – this is a typical example of a design-related dispute. I have included a commentary on the way it was settled, to show how it follows the general model of Fig. 9.11.

The radial fan case study dispute

Background
The case study conclusions, so far, have clearly identified the design (of the shaft/key slot arrangement) as the proximate cause of failure. Figure 9.6 was compiled as a failure/causation diagram to help explain how the failure mechanisms (categorization) linked in to the issue of causation. It was also identified that two other factors contributed to the failure but were not sufficient, in themselves, to be classed as the proximate cause. The implications of these conclusions are clear – the designer was at fault – and the contributory factors do little to mitigate against the fact that the design was wrong. It is also implied in the case study that manufacturing compliance was satisfactory, the design inadequacies being already included in the drawings and specifications before manufacturing started.

The questions
Following submission of the written case study report (Figs 9.5 to 9.7), and your initial verbal presentation, the following questions are put forward by the fan manufacturer:

'... looking at Fig 9.6, it seems to me that it was the operation, rather than the design that was the prime cause of failure' ... [*wrong terminology*]. He continued: '... so it was a maintenance problem, not recognizing that the impeller was out of balance, wasn't it?'

This was apparently backed up by someone else in the same company, but she started off on a different tack:

'... Don't you think it was a material defect problem? – I can't see how cracks can just 'happen', even if there is a bit of vibration' ... Now the thrust: 'there must have been a defect in the material, so why didn't you find it?'

The answers
The first question is really about the nature of fatigue, even though the questioner has not addressed it by name. It is also likely that he is

struggling with the meaning of proximate cause (he hasn't mentioned that either). Here is your answer:

'It's really a case of the way that the failure *started*.' You gather speed: 'The proximate cause is the cause that is responsible for setting in motion the chain of events leading to the failure – the *design* did that. Also, the vibration alone would not have caused the failure, the fan would still be operating normally if the crack hadn't initiated.' You open up the definition a little further: '... proximate refers to a cause which is *proximate in efficiency*, meaning the thing that has the greatest *effect*.'

The second question is fundamentally a metallurgical one. You are confident that neither the macro- nor micro-analysis show any identifiable 'defect'. You answer:

'I'm afraid that fatigue *can* cause its own crack initiation. If the stress concentration conditions are sufficient, then normal levels of operating stress can start a crack – it doesn't need an inherent defect to help it. This is accepted metallurgical theory ...' (well, one of them, anyway) '... which means that we come back to design as being the proximate cause.'

Further technical analysis

In view of the various reservations in acceptance of your condemnation of the design, some further technical analysis is advisable to reinforce your position. This is focused around the issue of demonstrating quantitatively that the 'as designed' keyway slot will cause excessive stress concentrations. You present these in written form, as an addition to your failure investigation report.

Figures 9.12(a) and (b) show one possible approach. Here, the existence of stress concentration is related to the 'fatigue strength' of the fan shaft. The principle is that the published principal yield stress (σ_y) is effectively *reduced* by three geometrical effects, experienced with the 'as designed' arrangement of the fan shaft, with its sharp-edged keyway slot and changes of section (shown in Fig. 6.8). When the shaft is subject to torsional fatigue, the actual torsional (shear) stress needed to cause the material to yield, and a crack to propagate, is only about 30 percent of that assumed if 'principal stress' σ_y assumptions are used. The coefficients shown are based on robust empirical data but are, strangely, rarely mentioned in design standards for general engineering equipment.

SUBJECT: TORSIONAL FATIGUE STRENGTH OF A STEPPED SHAFT CONTAINING A KEY SLOT.

1. Size effect
It is well established that fatigue strength of a shaft reduces as its diameter increases because of:

- increased likelihood of surface defects (more surface area)
- greater residual stresses (owing to the thicker material section) giving a more uneven stress distribution.

Typical handbook data gives the effect on torsional fatigue limit as a coefficient (C). The difference between the fatigue limit of a 15 mm diameter test specimen and a 100 mm shaft is given as approximately 40% (reference source).

2. Surface finish
Rough surface finish decreases significantly the fatigue strength of shafts subject to cyclic torsional loading. Typical data are:

Finish	Coefficient
Rough turned 2.5 µm R_a (N11)	1.80
Finish turned 3.2 µm R_a (N6)	1.40
Medium ground 0.1 µm R_a (N3)	1.10
Fine ground 0.05 µm R_a (N2)	1.00

Figure 9.12(a)   Further technical analysis – fatigue limit

More discussions
Following your further technical submissions the next meeting is attended by a reduced number of people (the others think the failure is already solved, or have decided that they don't want to be around in case the problem is somehow deemed to be 'their fault'). You present your further technical points verbally, and are progressing, logically and incrementally, towards explaining the compounded stress concentration effect of sharp keyway corners and rough surface finish using the table in Fig. 9.12(b).

'Just hold on – we've been making keyway slots like that for years, are you telling me how to manufacture my own fan? I'm the

3. Notch effect

There is a decrease in fatigue strength owing to both the sharp-edged key slot and its proximity to the shaft step. This is expressed as a coefficient. Typical data are given below:

Geometrical arrangement	Coefficient
Plane shaft, no keyway	1.00
Plane shaft, with 3 mm radius keyway	1.15
Stepped shaft with 3 mm radius keyway	1.32
Stepped shaft with 3 mm sharp-edged keyway	1.67

SUMMARY

The combined effects of these factors are:

'Ideal' fatigue strength	$= \sigma_y$
Size effect	$= \times 1.4$
Surface finish effect	$= \times 1.4$
Keyway notch effect	$= \times 1.67$
Cumulative factor for fatigue strength	$=$ 3.27

Hence the fatigue strength of the component is reduced to about 30% of its 'rated' strength value.

Figure 9.12(b)   Further technical analysis – fatigue limit

manufacturer here, not you, twenty five years at this game, man and boy ... I was just saying to my MD the other day ...'

You don't need to interrupt this chap yet, he's not going anywhere useful. He's the manufacturer anyway, not the designer. He continues:

'... keyway slots are always sharp, you can't buy round edged milling tools, and surface finishes below 12.5 µm R_a put the cost right up and ...'

Remember the guidelines: keep listening, keep answering, and be careful not to introduce anything technically new. Just amplify what you have concluded already. These types of questions are not difficult to answer factually if necessary – Fig. 6.9 from the 'Design Assessment' chapter shows the standard references for dimensional and surface tolerances on keyway slots.

The settlement – commentary

The dispute so far has followed the general pattern of Fig. 9.11. The initial design assessment has answered most of the technical questions, and has been supplemented by a further, controlled level of technical analysis to reinforce further the pointers towards *design* being the proximate cause of the failure. The last series of questions, which are fairly typical, lack firm direction and can be answered factually, leaving only the final comparison of the shaft/key slot design of other similar shaft couplings to be introduced as the final precedent. The technical standards also provide firm evidence that the existing design is inadequate for the purpose for which it was intended. In line with Fig. 9.11, I have tried to show how a small number of technical issues (perhaps 20 percent) inevitably remain floating and can often not be completely defined as adequate or inadequate, however long the dispute goes on.

Disputes have to end somewhere. A little understanding of the pattern that they follow can help, but you still need good engineering knowledge to back it up. In so many ways, settlement of failure investigation disputes is a function of the technical *confidence* of the parties involved. Straightforward lack of confidence (and sometimes competence) is perhaps the biggest factor in causing delays in the settlement of failure disputes. This is so well proven as to be almost an engineering *law*.

KEY POINT SUMMARY: PRESENTING YOUR CONCLUSIONS

1. Conclusions

The technical presentation of conclusions falls neatly into three parts:

- the formal (written) failure report
- your verbal presentation
- the dispute.

2. The written report

- This is based around the structure and content of the causation statement (see Chapter 8).
- It includes a failure/causation summary *diagram*.
- An executive summary is essential to make things clear.

3. Your verbal presentation

- This is your chance to *present* the contents of your written report.
- It should follow the content of the report and be predictable, with no hidden surprises.
- Expect *questions*.

4. The dispute

Many failure investigations end up in a dispute. This will involve the failure investigator (you) in:

- extra bits of analysis and reporting
- answering technical questions (some harder than others)
- discussions and meetings
- more discussions and even more meetings.

There is a certain 'routine' to most disputes – but they all have to end, sometime.

Chapter 10

Insurance investigations

Many engineering failure investigations are driven by the commercial and working practices and ways of the business world. A high proportion, perhaps more than half, are commissioned to help with the settlement of insurance claims – the type A investigation introduced at the very beginning of Part I of this book. If you include those investigations that follow a similar rationale to insurance-related investigations, i.e. commercial and liability cases, then the percentage rises. It is fair to conclude that 80–85 percent of the investigations in which you will be involved will have, at their root, a similar set of commercial objectives. The most common type is the straightforward insurance investigation in which an engineering failure has occurred, leading to a potential insurance claim. Within this scenario there are several different technical investigation roles – in the UK and USA it is possible for an investigation to be commissioned by the party making the claim (the insured) or by the insurers, or the appointed loss adjusters. Continental Europe has a slightly different system, using appointed 'experts'. The *technical* form of the investigation, however, is much the same, even though it may differ slightly in its detail.

The format of insurance-related investigations is fairly well defined, mainly by precedent ('it has evolved that way'). They have their own particular set of terminology which can be confusing if you are not familiar with it – not all the terms mean the same as their common English usage. The reporting format is simpler than you may expect, but it is highly focused, concentrating only on those aspects which are of importance in settlement of the insurance claim. This final chapter of the book concentrates on such insurance-related failure investigations. Combined with the more technically based information in the other chapters it should enable you to take part with confidence in these investigations. We will develop the technique first, in stages, and then see how it is applied to the radial fan case study.

The principles of insurance

If you were to search long and hard, you would eventually find, probably on a mountain somewhere, the definitive version of the principles of insurance. They would not of course be written on paper (they have taken several hundreds of years to develop) – stone is more likely. Showing evidence of the work of a thousand stonemasons, the principles would be expressed *separately*, so you would have to step back several kilometres (being careful not to fall off the mountain) to see them in context (Fig. 10.1). The carved rules would tell you a story – a consistent story, undeniably – but you could be excused for thinking that they represented a rather strange way of looking at the world. On your journey back down the (slippery) slope, resplendent in mild indignance, you pass the suggestions box inscribed 'any better ideas?' Sadly, you find it empty.

Despite their (sometimes) intangibility, these principles of insurance are the driving force behind so many engineering failure investigations. While there is no need to have knowledge of their detail, it is essential to have a basic understanding of the *territory* of insurance. This will give you a basic appreciation of how its principles influence the engineering failure investigation process. Figure 10.1 shows six of the principles – we can look at them in broad order of importance.

Indemnity

The idea of indemnity is one of the basic concepts of insurance. Indemnity means returning the insured party to the same financial position after a failure (loss) as enjoyed before the loss or, put more simply:

> Indemnity is an exact financial *compensation*.

This is very relevant to the engineering world where an equipment user insures against failure of the equipment, and possibly also the financial consequences of it being out of use. Indemnity is a well-proven principle and is supported by common law. The key fact is as shown above; indemnity extends only as far as exact financial compensation – it does not pretend to improve someone's financial position over and above that of before the loss occurred. You will see this has important implications for the process of putting failures right, without improving the original design.

Insured perils

A peril is the formal name given to an event or happening that results in

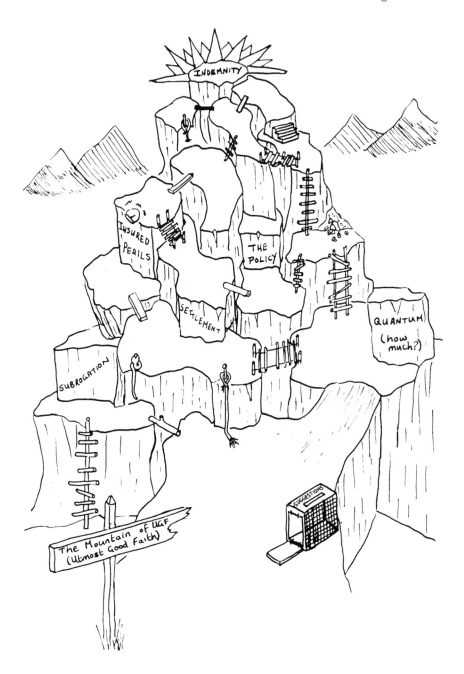

Figure 10.1   Genesis (well, not quite) – the principles of insurance

a loss. Traditional perils are fire, flood, theft, etc. or, in an engineering context, inadequate design, mal-operation (a type of negligence) or various forms of external factors such as impact or accident. Insurance policies are written around the idea of insured perils – perils which are recognized as being covered under the policy of insurance. The other important category is the *excepted peril*, a peril which is specifically excluded from insurance cover because it is not practical to insure against it. Excepted perils are normally mentioned in the insurance policy. The final category is *uninsured perils* – perils which are clearly outside the scope of an insurance policy but not mentioned in the policy document. Figure 10.2 summarizes the situation.

The policy

Everyone knows that this is the written document setting out the conditions of insurance, and saying what is covered. There are a few clever names given to the different types – you may see reference to 'narrative-style', 'modern language' and 'scheduled policies' – but they are essentially only different ways of expressing the same thing; the terms and conditions of the insurance contract. Don't expect policy wording to be exciting – it is meant to resound with a touch of gravitas. Interestingly, under another set of historical precedents, the policy document is actually *not* the contract, but merely written evidence of it. Practically, however, the policy document itself is generally taken as

An *insured* peril	A peril which is recognized as covered by the policy of insurance. Examples are fire and theft.
An *uninsured* peril	A peril clearly outside the scope of the insurance policy and which is not mentioned in it.
An *excepted* peril	A peril which is specifically excluded from the policy cover. Examples are: war, riot, and some types of nuclear and chemical accidents.

A peril is an event or happening, which can result in a loss. A peril is a happening, not a condition – so negligence, for instance, is not necessarily a peril.

Figure 10.2   The three types of 'perils'

being 'the intention of the parties' (insurer and insured), in the absence of proof of the contrary. You will find that the content of the policy is an issue in many failure investigations. While the insurers, insured, and loss adjuster pore over it there are some points to note from your more technical viewpoint:

- *Terminology.* The main categories of insured perils do not fit neatly with the categories of failure that we saw in Chapter 7. The insurance policy will refer to perils in a much more general way. Do not expect to find, for instance, reference to specific *technical terms* such as principal stress failure, fatigue, or creep. You are more likely to find only rather simple statements saying that the insured is covered for 'sudden and unforeseen damage' or 'breakdown'.
- *Equipment design and operation.* Again, you won't always find these mentioned. Inadequate design and/or incorrect operation may be covered under the guise of negligence which, although strictly not classed as a peril, *is* insurable.

By now you have maybe concluded that while the wording of the policy may be a source of fascination to other people, it doesn't hold a lot of interest for the technical investigator. It isn't going to affect the content of your technical conclusions on failure categorization and causation – although you can expect it to have an influence on the way that your conclusions will be interpreted and used.

Settlement
If the insurance policy wording is the hushed pianissimo of the music heard on the mountain, then the term *settlement* is the final crescendo – trumpets playing and drums pounding. During the first meeting held between the insured and their insurers following a failure, both sides will be heard proclaiming loudly that their only objective is to pay homage to the great god of settlement. Sometimes they may even *mean* it. Soon, however, the music dies down – and it is back to the mechanics of the investigation, with its repetitive chorus of 'proximate cause'.

Settlement is about money of some sort. Insurance terminology doesn't put it quite like that, of course – instead it is referred to in the context of the *provision* of the indemnity. The method of this provision is written into the policy but the general term used is *reinstatement*. There are three different methods available: cash settlement, repair, or replacement.

Cash settlement
This is the most obvious, and in some ways the most suitable, method of

settling insurance claims for damage or failed equipment. Many claims are settled in this way. It is often only an *option* under the policy, however, and the majority of claims are settled in other ways.

Repair
When equipment has suffered a failure but is still repairable, then adequate repair is an adequate way to fulfil the indemnity. It is less applicable to industrial equipment claims than, for example, to motor insurance.

Replacement
Replacement is a practical method of reinstatement when the loss includes small or delicate pieces of equipment which are easily replaceable. For large capital items, which will have degenerated either financially (by deterioration) or technically (wear and tear) prior to the failure, it is not a practical way of settlement – so you will rarely see it written as a policy condition. It finds common use in 'new for old' household and domestic goods policies.

The technical chapters of this book emphasize that settlement is perhaps the primary objective of an engineering failure investigation. From a technical perspective, it is easy to confuse settlement with the attainment of absolute technical agreement between the insured and insurer on the nature of failure. I have explained previously that this is not necessarily true, so you shouldn't pursue it as an ideal. Settlement means *settlement* – not necessarily agreement.

Quantum
This is a piece of classical insurance-speak:

QUANTUM

means

'OK, we agree that there may be a valid claim, but how much?'

We are getting down to fine (but rather academic) insurance definitions here. Insurers like to keep those discussions about whether they have a *liability* under the policy separate from the subsequent 'inquisition' about *how* the claim will be settled. This differentiation of the 'is it?' from the 'OK, how much?' becomes rather time-consuming, and occasionally amusing, but that is how it is done. For this reason, the term quantum tends to be used a lot, even protruding into the technical discussions when the investigation starts to impinge upon the financial implications of the repair or reinstatement of the damage. You don't need to get too worried by the terminology – simply remember that 'quantum' means 'how much?' or perhaps 'how much if?'

Subrogation

'Subrogation' is the term guaranteed to bring a glint to the eye of the pragmatic failure investigator, frustrated by the apparent gentleness and decorum of all the other principles of insurance we have looked at so far. This is the move from the defence of the drawing room to the offence of the rock face – giving some *point* to the exercise. Subrogation is a key part of the territory of insurance investigations. It is also one of the activities of *direct relevance* to the technical activities of failure investigation. You could be excused for concluding that in the world of effective failure investigation, the word 'subrogation' lurks behind every piece of broken metal.

Subrogation – what is it?
The word comes from two separate Latin words; *sub* (under) and *rogare* (to ask). Hence subrogation means 'asking for (a payment) under someone else's name'. The most common context is where the insurers exercise the rights to recover money from a third party in the name of the insured. Figure 10.3 shows the situation. Common law subrogation rights are acquired by insurers once they have provided their insured (policy holder) with 'the indemnity' that everyone has agreed has been

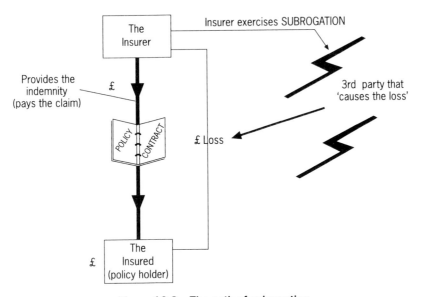

Figure 10.3   The path of subrogation

provided for by the insurance contract. There is normally a specific contract clause covering this. The third party is generally the party that has been deemed to be 'responsible' for the failure.

Is it legal?
Yes. The basic tenet of subrogation has long been accepted under common law. Insurers can exercise 'rights or remedies' (formal terminology again) possessed by the insured for their own benefit, i.e. to recover the money that they will pay the insured who has suffered the loss. There is a general 'commonsense' caveat that the common law acceptance of subrogation only applies *after* an indemnity has been provided, meaning that insurers cannot exercise subrogation rights if, put bluntly, they are not going to accept liability for the claim. In practice, this common law position is often amended by a policy clause, enabling the insurer to start proceedings before providing an indemnity.

Subrogation rights
The insurer can only claim money from a third party if that party has a liability, i.e. the insurer must have the *right of recovery*. There are four main ways that subrogation rights can arise. They are:

- *In tort.* A tort is a civil wrongdoing, such as negligence. If the proximate cause of failure can be clearly linked to the negligence of a third party, then the third party is said to be 'liable to the insured in tort' (note that they are not liable to the insurers, who will merely act in the *name* of the insured). Negligence may be linked to the design, manufacture, assembly, operation, maintenance, or any of the other 'real engineering' activities relevant to the failed equipment.
- *By contract.* If a third party has a contractual responsibility (such as a maintenance contract or hire agreement) to the insured which includes provision for the repair of damage etc. then that party may be liable to the insured to pay for or repair the damage, even if negligence was not involved. Straightforward 'subrogation under contract' is actually quite rare in engineering failure cases – it is much more common for the situation to revert to 'subrogation in tort' instead.
- *Under statute.* Don't worry about this one. It refers to the few cases where Acts of Parliament give subrogation rights which might not otherwise exist. The Sale of Goods Act is one example – it can be relevant to mechanical failures of engineering equipment.
- *Under salvage.* The term 'salvage' is used when a failure causes a 'total loss' of the equipment and the insured is paid an amount that

reflects this, i.e. no allowance is made for any monetary value that failed equipment may have left. The most obvious example is a wrecked motor vehicle but the same principle applies to industrial plant. Salvage is a rather shaky form of subrogation right and is not as well precedented as the other three. You shouldn't need to be involved in it very often – if at all.

The implications of subrogation
Subrogation is the principle of insurance that has perhaps the most influence on your technical investigation, particularly on the possible effects of the things that you say, do, and write. Because subrogation actions are always a possibility, it is important that they form part of your 'focus' on a failure investigation. Here is the key point:

During your investigation (and before and after it as well) you must not do, or omit, anything that will prejudice possible subrogation rights. This is part of your 'duty of care'.

Note the wording – it says *possible* subrogation rights. As technical investigator it is not your job to decide whether subrogation rights exist, or to edge the case gently towards, or away from, a subrogation action. It is, however, your responsibility not to do anything which will make successful subrogation more difficult, or impossible, should it happen. This has an important effect on the way that you formulate your technical report for an insurance-related failure. *Objectivity* is the key point – look back to Chapter 3 of this book if you need reminding further about the nature of objectivity.

These, then, are the six principles of insurance (at least as they apply to you). There are of course many more. You could spend the first half of the rest of your life learning about the others, leaving plenty of time to reflect, in the second half, whether you actually gained anything useful from the exercise. The deeper concepts and muted terminology are fascinating, but you are well advised to leave it to the insurers and lawyers. You will help enrich their lives.

This chapter started by introducing the concept that Fig. 10.1 represents the territory of insurance. All you need to know now is which land you are in, and you have the full picture. After all, if you don't know which land you are in, how do you know what language to talk? Look back at Fig. 10.1 – do you see the signpost? – you are in the 'land of UGF'. This idea of UGF is what governs so many things in the world of insurance claims – it seems that the word 'insurance' itself should perhaps be redefined as 'insurance (UGF)'.

UGF stands for 'utmost good faith'.

The principle of 'utmost good faith' is what differentiates insurance contracts from more traditional forms of commercial contract. In commercial contracts the common law principle is still 'caveat emptor' (let the buyer beware) – this implies that the parties involved are not required to reveal everything they know about the proposed agreement – they are subject only to *good faith*. Insurance contracts are different – the doctrine of caveat emptor does not apply. Instead, insurance contracts have a fiduciary nature; meaning they are based on mutual trust and confidence between the insurer and the insured. This goes under the less well known common law principle of utmost good faith, where the parties to an insurance contract are legally obliged to reveal all relevant information, whether it is requested or not. This includes technical information. This revelation of information happens mainly before the insurance contract is agreed upon.

After the contract has been finalized, i.e. during the period when a claim could be made, the position regarding UGF is less clear. It is sometimes argued that the situation regresses back to the traditional 'good faith', or alternatively that UGF still applies. You can imagine the horrendous rhetorical arguments that can (and do) arise out of this situation. Happily, the dilemma rarely shows itself. The atmosphere of insurance claims is nearly always one of UGF whether or not it is needed to comply with the strict legal implication of the insurance contract. Hence the activities of notification of claim, the technical investigation, and the settlement (that most important step, remember) tend to be carried out in a way that is reminiscent more of mutual co-operation than open hostility. You can expect to see the effects of this during the technical investigation, where good technical co-operation between all parties is the norm. This also places responsibility on you to co-operate technically and disclose relevant information to parties other than your immediate client. I have come to the conclusion that this is all for the better. There is little choice, as this is the way that the system seems to work. UGF may also apply to the subrogation activity as well. Draw your own conclusions – but read Part I of this book first.

The insurance investigation

The principles of insurance discussed earlier in this chapter are fine, but they are still only principles – the important point is the way in which they manifest themselves as influences on the failure investigation itself.

Insurance-related failure investigations are a particularly *focused* type of technical investigation, so it is worth looking in a little more detail at their structure. The most common type is that following the failure of a piece of mechanical equipment that is covered under an insurance policy providing an indemnity against 'machine breakdown' or 'all risks' and resulting business interruption (BI) loss.

Terms of reference

Insurance-related failure investigations should always have a set of explicit terms of reference (TOR). Although the basic objectives of the investigation don't vary much between cases, there can be differences in the detail, mainly caused by the content of the policy conditions. Even in traditional 'engineering insurance', there are quite a few different sets of policy conditions, explaining which perils are insured.

It is surprisingly rare for insurers, loss adjusters, or even insured policyholders to know exactly what to put in the TORs used when they commission the services of specialist failure investigators. This doesn't mean that your clients don't know what they *want* – they do – but it can often be difficult to express. The restricted terminology of the insurance world doesn't help – if insurers stated their requirements in strict insurance terminology it would not be understood by many consulting engineering companies or individual engineers that offer their services as failure investigators. You can see the danger in accepting an investigation case that has no explicit terms of reference – it is easy for you and your client to have different understandings about what the investigation is meant to achieve. Unfortunately, some insurance-related investigations end like this, with both parties feeling disappointed and the failure still not 'solved'.

There is no reason why TORs have to be complicated, or long-winded. It is important, however, that they are *clear*, and use the correct terminology. Figure 10.4 shows a typical set used for a material damage/business interruption policy claim where the insurer felt the need to get professional engineering advice on an equipment failure. Note how there is a mention of subrogation in the TOR. If there isn't, it does not mean that it is not going to be an issue as the claim progresses – rather it reflects on the reluctance of some people to mention it specifically. It is always in the background, but sometimes not written down. It is one of the stranger aspects of failure investigation that you can't ignore subrogation, just because it doesn't appear in the official TOR. This is due purely to convention – the world of insurance, remember, is a

---

TERMS OF REFERENCE

An important piece of rotating equipment in a chemical plant has failed. The insured has notified the insurers of a claim for material damage and business interruption. The insurers have commissioned a professional failure investigator to provide a written report and presentation. The terms of reference are:

1. Quantify the nature and extent of insurable damage.
2. Define the chronological sequence of events related to the failure.
3. Determine the proximate cause of the failure.
4. Report on the insured's technical proposal for repair/reinstatement of the equipment.
5. Report on any technical aspects that may arise in subrogation.

Figure 10.4 Simple TOR for an insurance-related failure investigation

deeply traditional land which places little value on attractive novelty, even if it does not hurt.

The parties involved

The structure of parties involved in insurance claims (particularly high-value ones) is different to that found in other parts of the commercial world. Insurance *risk* is bought and sold, rather like a commodity, with the result that it is rare for a single insurer to carry all the risk. The whole business goes under the term *reinsurance* – one of the basic tenets of the insurance world. It has two main effects: it spreads the risk between multiple insurers and, in doing so, makes the activities of assessing and paying claims more complicated. In some cases, the original insurer (or 'cedant') of a risk will only actually be liable for a few percent of the cost of a claim, with a hierarchy of reinsurers liable to pay the remainder. Fortunately, the whole thing falls into the simplified pattern shown in Fig. 10.5. The main players are:

- *Reinsurer(s)* generally have the right to 'control' a claim under specific clauses in their contract.
- *The insurer*, also known as the 'cedant' when there are reinsurers involved, is the party who provides the indemnity to the insured. Under the rules of insurance, the insured can only claim directly from their insurers and has no access to the reinsurers.
- *Loss adjuster(s)* may be appointed by either the insurers or the

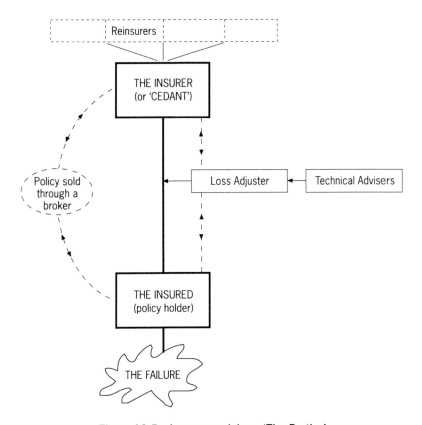

Figure 10.5   Insurance claim – 'The Parties'

reinsurers. Their role is to investigate the claim (the issues of liability, quantum, and subrogation) with the insured who is making the claim and then make recommendations to the insurer. The position of loss adjusters is difficult – under their charter they have a duty of care to both insurer and insured and so are required to act in a way which is independent and impartial when dealing with a claim. It is wrong to assume that they act only as the agent of the insurers. Loss adjusters provide the necessary 'bridge' between insurers and the details of a claim. They are also the party most likely to recommend a technical investigation into a failure, as they do not normally have formal engineering qualifications themselves.

- *The insured* is the party that pays the insurance policy premium in return for the indemnity. They often buy the policy through brokers.

This structure of parties is a little unusual – to an engineer the position of the brokers and loss adjusters might seem at times almost superfluous, given that a clear written contract (of insurance) exists between insurers and their policyholder. In practice, however, this is the way that the business operates; partly because it *does* work and maybe partly 'because it has always been done like this'. Historical precedent is an important consideration in the insurance world.

Your findings

The most logical way to organize the technical findings in an insurance-related failure investigation is to follow the structure of the terms of reference. The basic ideas of defining failure categorization and causation (Chapters 7 and 8) work well in this context. There are a few special characteristics to note.

Damage definitions

We looked at the technical definitions of 'damage' in Part I of this book and saw that it is capable of fairly firm definition in its engineering context. The situation is different in insurance investigations. One term you might see is *material damage* which, although well understood in an insurance context, is not easily cross-referenced to the technical definitions of damage discussed earlier. Within the context of 'material damage', it is still necessary to define, in engineering terms, ('categorization'), the type of damage that has occurred.

It is these differences in terminology that sometimes cause confusion in insurance-related failure investigation – if you are not careful, everyone can end up talking about different types of 'damage'.

Events

The chronology of events is of critical importance in insurance claims. You will see the term used a lot in letters passing between insurers and reinsurers. Strangely, you will have difficulty in locating a formal insurance definition of 'an event'. The normal understanding in insurance investigations is its common-usage interpretation: 'the event' is the happening that *initiated* the sequence of happenings that resulted in the sudden unforeseen failure.

There is less misunderstanding about this definition than you might think and it is rare for insurers to stand on strict technicality as long as 'the event' clearly occurred during the period of insurance cover. Figure 10.6 shows how the analysis of chronology works in practice.

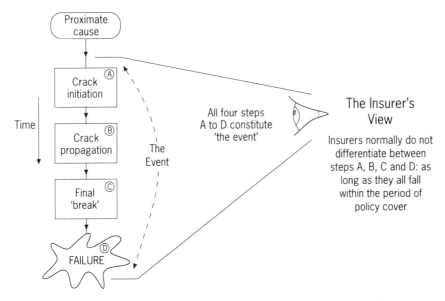

Figure 10.6 An insurer's understanding of 'the failure event'

Proximate cause
All the information on proximate cause provided earlier in this book applies to insurance-related failure investigations. The guideline is simple: find it, then say what it is.

Repair/reinstatement
As technical adviser, you may be more often asked to *review* technical options for repair or reinstatement of failed equipment rather than to propose what they should be. Checking of costs is often also involved. From an engineering viewpoint, one of the more challenging parts of this exercise is to make an assessment of betterment. When equipment is replaced or reinstated, there is nearly always some change from the original pre-failure specification. Even simple equipment design develops quite quickly as various parts are changed and upgraded. High technology products such as gas turbines and complex process equipment are the best illustration of this – they are technically obsolete almost as soon as they are sold and certainly by the time a major failure occurs. Assessing betterment is such a varied activity that it is difficult to give firm guidelines that will be relevant in all cases. The main thing is to *make it clear* – if you find a clear technical reason why the repair/ reinstatement improves the engineering performance of the equipment then say so. There are three common sources of betterment:

- better *performance* (greater output, or improved efficiency)
- longer lifetime
- improved reliability.

It is often difficult to separate them as clearly as you would like, but you can at least make an attempt. Use quantitative values (percentages, lifetimes, output levels, etc.) wherever possible. This will make the situation easier to understand for all parties concerned – there may be large sums of money resting on the interpretation of your technical statements on betterment.

Reporting format

There is no single reporting format that is suitable for every insurance-related failure investigation you are likely to meet. There are too many sets of different circumstances and variations in insurance policy conditions for such a level of prescription. Conversely, there is a danger that a reporting format which is left too loose will not properly address the issues. There is also the issue of presentation – a failure investigation can involve a lot of painstaking work, so it is natural to want to present the results in a way that your clients will find useful. Most insurers and loss adjusters like a technical investigation report to be presented in two parts (Fig. 10.7).

- first, your *preliminary advice*
                                    then
- your detailed report.

Your preliminary advice

The purpose of preliminary advice is to inform the insurer or loss adjuster, as soon as possible, of the preliminary technical facts of the case. The need is driven by the requirement for the loss adjuster or direct insurer (cedant) to 'place the parties on notice'. This means simply that the insurers and reinsurers, as soon as possible after the failure has occurred, like to be informed about the possible nature and size of any future financial liability (i.e. the claim). This knowledge is needed to allocate an informed 'reserve value' to the claim. The actual sum allocated is usually conservative so it can be amended gradually as the investigation progresses and the technical details, and financial consequences, become clearer.

The format of the preliminary advice is not unlike that of the executive summary of the subsequent main technical report, but it is simpler. Your client will understand, and accept, that the advice is your 'best estimate' made in advance of your full investigation, and so may be

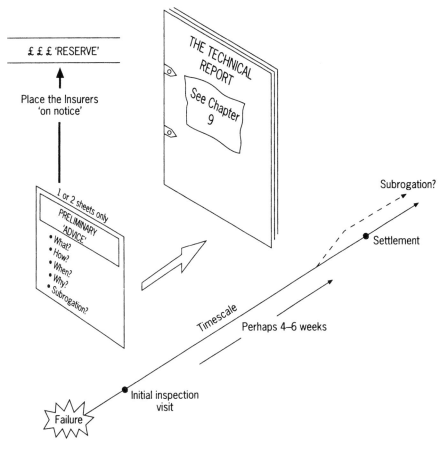

Figure 10.7 Reporting format – insurance investigations

subject to change. This means that there is no need to be indecisive or (worse) evasive in your preliminary advice – the objective is to say what you think, using the best information available at the time. Salient points to try to follow are:

- *Length*: a preliminary advice report should ideally occupy a maximum of two pages of text; a single page is better. Diagrams or photographs are not normally used at this stage.
- *Format*: a simple list of chronological events or 'bullet points' makes things easier to understand.
- *Content*: in general, follow the same content structure that you will use in the final causation statement (shown previously in Fig. 8.6)
  – Identify the failed equipment.

- Mention how it was operated.
- State the circumstances of the damage or failure.
- Define the 'nature and extent' of damage.
- Give a broad indication as to what you *think* may be the cause of the failure, but don't try to define proximate cause yet.
- Mention if any third party may have been involved. This is to give an early indication of any potential subrogation possibilities.

Done correctly, the preliminary advice is an important element in maintaining the efficiency of insurance-related failure investigations. Some of the simpler claims can actually be settled on the basis of a preliminary advice report, eliminating the need for further expensive investigation work. This is absolute good practice – you will not gain anything by extending investigation work beyond the point where it needs to go to act as an 'enabler' for the settlement of the claim.

Your technical report
Follow all the general guidance on content and format given in Chapter 9. The only real difference between an insurance-specific report and one suitable for a more general technical or commercial investigation is in the use of terminology. Terms such as 'proximate cause' and failure 'event' have specific meaning in the insurance world, so it is important to use them with the same *precision* as they will be interpreted. In some cases it is possible to strip out quite a lot of the technical detail from the report without lessening its value to your insurer or loss adjuster client – just be careful not to go too far. If in doubt, it is probably better to leave the technical detail in. Without doubt, the main objective of an insurance-specific failure investigation report is to keep a good sharp focus on the subject of *causation*. Technical details and arguments are fine, but it is reports which point clearly and unambiguously to the *cause of failure* which will hold your client's interest in your report (and you).

Back to the case study

The case study 'story' so far is fairly typical of what you could experience when asked to investigate a machinery failure. What we have seen is based broadly on an insurance-related investigation, mainly because insurance claims are one of the primary driving forces behind engineering failure investigations. Note the steps that we have gone through already for the radial fan case. We have:

• Described, technically, the equipment and its mode of operation.

- Compiled a basic failure/causation diagram (Fig. 9.6) to help simplify the explanation of the failure events.
- Defined the proximate cause of failure, and its surrounding context of contributory factors.
- Considered the subrogation possibilities, and have not said or reported anything that would compromise the possibility of a successful subrogation action.

Dispute

Most of the work is done. The only item of any significance left is the business of planning for the forthcoming dispute (we discussed the premise that while there is not *guaranteed* to be a dispute, if you plan for it you will rarely be disappointed – controversial maybe, but true). The radial fan case study is a useful example with which to look briefly at some facets of the possible dispute, in the way that it *could* develop. It is difficult to be very specific about the outcome of these parts of the dispute, mainly because of possible variations in the written conditions of the applicable insurance policy – and there are good reasons why I cannot make the case study policy-specific. We can, however, look at three of the most common areas of dispute that could develop in this fan failure. For this we need to revert back to the previous failure/causation diagram, shown again as Fig. 10.8.

The definition of events
This is the most likely area of dispute to arise, but is also one of the easier ones to solve. Figure 10.8 (and Fig. 10.6) clearly shows the event as being comprised of crack initiation, propagation and then final ductile fracture as the shaft breaks. Here is a contrary argument:

'... the event was not the breakage of the shaft, it was the crack initiation ... and it wasn't the initiation that caused the shaft to break ...'

This is more the result of misunderstanding than mal-intent. From the description of the failure initiation and propagation in Fig. 10.8 and the section on fatigue progression in Chapter 7 it should be clear that there is *continuity* in the way that the crack grows from its initiation phase to the point where it provides the condition for the shaft to fail. The key fact here is that the idea of continuity infers an *unbroken sequence of events*.

The concept of an unbroken (and continuous) sequence of events is fundamental to the way in which 'events' are interpreted in insurance investigations. By 'unbroken' it means that there has been no chance for

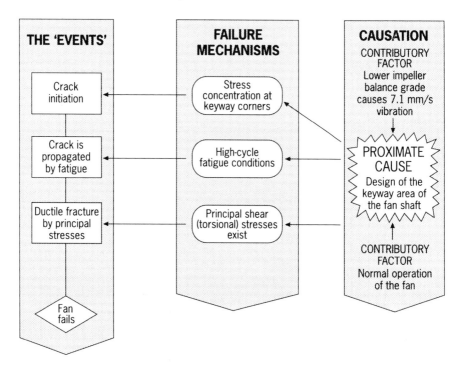

Note how the events, mechanisms and 'causation' are expressed separately but the linkages between them are clearly shown

Figure 10.8   The radial fan case study – failure causation diagram

some other (possibly undefined) outside influence to have helped the failure. The outside influences of particular interest to the insurance policy are those which come from an excepted or uninsured peril. If this *had* happened, there might be a good prima facie argument for the claim being invalid and someone would have to look deeper into the policy conditions. Can you see now why precise descriptions of failure mechanisms are so important? Fortunately, for this radial fan case, the continuity of the mechanism, as described, qualifies as the 'unbroken sequence of events' – so the exact definition of the event as either initiation, propagation, then failure, or simply initiation only becomes of academic interest. Insurers look at each case on its merits.

The definition of proximate cause
Expect disputes about proximate cause. You could expect to hear something like:

'... the shaft *design* can't be the proximate cause ... it was the impeller imbalance that caused the fatigue, so why isn't that the proximate cause ...?'

Proximate cause is often *thought* to be open to well-intentioned interpretation. While it is possible that there cannot always be a unique definition of it in every individual failure case, it is also true that it is more robust than many other terms used in engineering and commerce. This is because:

- proximate cause is well defined

and

- it has been heavily precedented during many thousands of insurance claim disputes. The modern definition has been well used and tested since a landmark legal case in 1907 (the case of *Pawsey* v *Scottish Union and National*, 1907).

The observation about whether it is the impeller imbalance that is the true proximate cause is not entirely unreasonable. The failure/causation diagram (Fig. 10.8) accepts that the impeller imbalance will cause increased vibration which in turn will result in higher bending and torsional cyclic stresses (i.e. the fatigue conditions). This is, however, not the *total* cause of the fatigue condition; the normal operating condition would also cause some stresses of the same type, so it is the cumulative effect of the normal and abnormal conditions that is the contributory factor to the failure. This viewpoint is also useful in reinforcing the semantic definition of proximate cause, the inference of proximity or *nearness*. It clearly points to the cause of failure which is proximate *in efficiency* in causing the failure. Hence the design of the shaft, with its sharp edged, rough-surfaced key slot situated near the change in section, is the cause which fulfils the definition. It also qualifies for the rest of the definition: the cause which 'sets in motion the chain of events ...' In most cases (as in this one) the true proximate cause can be identified clearly, if approached in the correct way.

Isn't the keyway slot classed as an 'inherent defect'?
This is another common one. Insurance claim investigations often start off under the premise that the failed item contained an inherent defect, perhaps conveniently shifting the blame from the equipment designer to its manufacturer. Again, it has a glimmer of believability – this time because of the weakness of metallurgical theory (Chapter 7 discussed how there is one, quite valid, theory of failure that assumes all metal is pre-cracked and all that cracks have to do to cause failure is to find the

right condition to allow them to propagate). The weak point in this argument is found in the concept of a case being proximate *in efficiency*. Given that, then as an extension to the 'always pre-cracked' theory it would be reasonable to expect multiple failures of fan shafts, whether they contained keyways or not. There is also the fact that the shaft failed at the keyway, and not somewhere else. This argument is less than absolute – it can become circular – so the best way is to revert back to precedent. Case precedent on inherent defects tends to reinforce the idea of an 'inherent defect' as an identifiable *manufacturing* (or assembly) defect such as the inclusions, porosity or similar findings discussed in Chapter 7. Once again, it is the accuracy of the technical *categorization* of the failure mechanism that holds your technical arguments together during difficult parts of the technical dispute. Loose definitions mean weak technical arguments – you will then need all the utmost good faith that you can get.

At last – the settlement
There is absolutely no logical reason why the radial fan failure case, if it was part of a valid insurance claim, should not be settled quickly to the satisfaction of both insurers and their insured. It is typical of many rotating machinery failure claims that you will see. The exact nature of the liability would depend on the policy wording; it could be covered by either a 'breakdown' or 'sudden and unforeseen damage' clause, both being used in engineering policies from major insurers. If you see failure cases like this one leading towards legal proceedings between insurer and insured, then the investigation process has gone wrong, somewhere.

And the road to subrogation?
It is unlikely that this case study would go to subrogation. While the proximate cause may be identified as being the *design* of the fan shaft, there are also contributory findings which could quite rightly be classed as 'mitigating factors'. The insurer would have the burden of *proving* that the fan was *not* fit for purpose – not such an easy task as it sounds. Coupled with this would be the probability that the financial claim on the designers, if successful, would be fairly small. Subrogation normally allows recoup of the cost of material damage only (new fan parts), but not usually the cost of business interruption (BI) losses. In a real insurance situation the BI losses on this case study would probably outweigh the material damage loss by a factor of twenty or more. Insurers do vary in their outlook, but for most I think the risk/reward profile of subrogation action in this case would prove unattractive.

KEY POINT SUMMARY: INSURANCE INVESTIGATIONS

A large proportion of engineering failure investigations are driven by *insurance-related* issues.

1. Insurance principles

The main principles of insurance that you need to know about are:

- indemnity (exact financial compensation for a loss)
- insured perils (happenings that result in a loss)
- the insurance policy (which sets out the 'context' of insurance)
- settlement (either in cash, or by repair, re-instatement or replacement)
- quantum (which simply means *how much?*)
- subrogation (the act of claiming financial compensation in the name of the insured party)
- UGF (meaning 'utmost good faith' – the philosophy behind all contracts in the insurance world).

2. The insurance investigation

To perform an effective insurance investigation you need:

- Clear terms of reference (TORs).
- An understanding of the roles of all the parties: policyholder, reinsurers, the insurer or 'cedant', loss adjusters, and technical advisers.
- Two stages of reporting, i.e.
  - your 'preliminary advice'

  followed by

  - your technical report (using correct terminology).

Conclusion

Engineering failure investigation is an interesting subject, but by no means an easy one. If you have worked through the chapters of this book it should be clear how it can be a complex (and sometimes a little uncomfortable) mixture of disciplines. Engineering and metallurgical knowledge are only part of the story – they have to be combined with an appreciation of the procedural, commercial, and insurance issues that surround most failure investigations. There is nothing unique about this. Many areas of engineering are, in reality, a similarly complex mixture of disciplines.

Whatever the surrounding context of a failure investigation, good engineering knowledge remains the most important part. The tasks of categorizing the failure mechanisms and translating them into precise descriptions of causation are purely technical – they cannot be done properly if you don't have enough knowledge of mechanics and metallurgy. An appreciation of mechanical design is also important.

We have seen, throughout the book, something of the many possible variations that engineering failure investigations can take. They are always a challenge – because their variety is so wide. Luckily, they do fall into patterns so you can generally be assured that there are some technical areas with which you will be familiar.

Investigating a failure is always a pro-active activity – you have to go looking for the answers, because (as we have seen) they are unlikely to come looking for you. Perhaps the most important message of the book is that failure investigations need good *conclusions*, and that there is little room for hedging or indecision. Your conclusions are the *product* of the business of investigation – ultimately, what your clients are paying for.

Parts I and II of this book have covered the main technical and procedural aspects of the subject. The boundaries between these two parts are not always as neat and discrete as you might like, so both parts are needed to get the full picture. Remember also that the information in this book, in itself, forms only part of what is a much wider discipline; technical information, particularly on metallurgy, is just as useful – as long as it is used in the correct context. I have included some further technical references in the bibliography – all competent books in their own subjects which can help your knowledge.

Clifford Matthews BSc, CEng, MBA

Meanwhile, back in the office

'What did that last chapter mean?'

'The bit on insurance investigations? It was trying to persuade you out of the idea that the cause of a failure can't be properly defined, to demonstrate that it isn't a case of infinite shades of grey'.

'Hmm ... it's just that this proximate cause thing seems a little bit, well, ... black and white for me'.

'Well, that's you, but ...'.

'I'd be happier if you could refer to 'possible causes', and the 'balance of probability' – things like that – I don't see why you can't'.

'Because you need a *label*, otherwise the technical conclusions have meaning only to yourself'.

'OK, then, I'll have several labels (I can name them after the failure categories in Chapter 7) and I can stick more than one on each failure – that way I won't ever be wrong'.

'No, you need selection – it's selection that brings engineering *precision* – and it's precision that makes the piston slide, and the shaft turn'.

'We'd need some strategy behind the labelling then?'

'Maybe that was what Part I was all about – and we both thought that all it was trying to do was to push us outside our technical comfort zones ... of course I spotted this, I ...'

'I like talking strategy – it's just unfortunate that you were slow to realize that the 'best strategy' must include the stability of its own results'.

'Er ... yes ... insurance investigations need their own type of strategy, it's the super-technical parts that ...'.

'You're not telling me you think that subrogation is a worthy cause ... for engineers that is ... like us?'

'It could be that subrogation is the art-form of the failure investigation process. Look at that Picasso water-colour in the boardroom – Picasso said that his art was a lie that showed you the truth – he probably thought the same about subrogation as well'.

'Try telling that to the MD, he only bought it to be decorative'

'So much for self-awareness ...'.

'Oh you need a lot of that for insurance investigations'.

'Awareness – of yourself and your technical facts?'

'Yes, and the techniques'.

'And the technical standards'.

'And the modes of failure'.

'And causation?'

'Always causation'.

'And then you have to package it all together, with an irrefutable faith that you are going to win, that there is absolutely no chance that you will lose ... Hey, where are you going?'

References

(1) *Kempe's engineers year-book* (Published annually) M G Information Services Limited, Tonbridge, UK.

(2) **Avallone, A. E.** and **Baumeister, T.**, 1996, *Marks' standard handbook for mechanical engineers*; McGraw Hill, New York, USA.

(3) BS 4500: 1985. *ISO Limits and fits*, Part 4500A, British Standards Institution, UK.

(4) BS 4235 Part 1: 1986. *Parallel and taper keys*, British Standards Institution, UK.

(5) ISO 1940–1: 1986 (identical to BS 6861 Part 1: 1987). *Balance quality requirements of rigid rotors: Part 1 – Method for determination of residual unbalance.*

(6) ISO 10816–1: 1995 (identical to BS 7854 Part 1: 1996 *Mechanical vibration, general guidelines*).

(7) BS 970 Part 3: 1991. *Bright bars for general engineering purposes*, British Standards Institution, UK.

Bibliography

Ross, B., 1995, *Investigating mechanical failures*, Chapman and Hall, London, UK.

Jones, D. R. H., 1993, *Materials failure analysis*, Pergamon Press, London, UK.

Meguid, S. A., 1989, *Engineering fracture mechanics*, Elsevier Science Limited, Barking, UK.

Nishida, S.-I., 1992, *Failure analysis in engineering applications*, Butterworth-Heinemann, UK.

Index